S0-ACR-482

FOOD POLIT

FOOD POLITICS
THE REGIONAL CONFLICT

EDITED BY

David N. Balaam

AND

Michael J. Carey

ALLANHELD, OSMUN   Publishers

ALBRIGHT COLLEGE LIBRARY

To Doobie and my family
To Barbara and my parents

Published in the United States of America in 1981
by Allanheld, Osmun & Co. Publishers, Inc.
(A Division of Littlefield, Adams & Company)
81 Adams Drive Totowa, New Jersey 07512
ISBN 0-916672-52-2

Copyright © 1981 by D. N. Balaam and Michael J. Carey

All rights reserved. No part of this publication may
be reproduced, stored in a retrieval system, or
transmitted in any form or by any means, electronic,
mechanical, photocopying, recording, or otherwise,
without the prior permission of the publisher.

Library of Congress Cataloging in Publishing Data
Main entry under title:

Food politics.

   Includes bibliographical references and index.
   1. Food supply—Addresses, essays, lectures.
2. Nutrition policy—Addresses, essays, lectures.
I. Balaam, David N., 1950-   II. Carey,
Michael, 1947-
HD9000.5.F596   338.1'9   79-48097
ISBN 0-916672-52-2   AACR2

Printed in the United States of America

338.19
F686C

183819

Contents

Tables and Figures   vii
Acknowledgments   ix

1   Introduction: The Political Economy of Food—the Regional
    Approach   1
    Michael J. Carey

PART I THE DEVELOPED REGIONS   9

2   Food Policy in North America: The Bread Basket   11
    Robert E. Dickens and Richard K. Moore

3   European Food Policy: Rules of the Game   30
    Michael J. Carey

4   Agri-Policy in the Soviet Union and Eastern Europe   48
    David N. Balaam and Michael J. Carey

PART II THE DEVELOPING REGIONS   81

5   Regionalism, Food Policy, and Domestic Political Economy:
    Lessons from Latin America   83
    Richard Ganzel

6   Asian Food Systems: Structural Constraints, Political Arenas,
    and Appropriate Food Strategies   106
    David N. Balaam

v

7   Agricultural Constraints and Bureaucratic Politics in
    the Middle East   143
      Marvin G. Weinbaum

8   Tropical Africa: Food or Famine?   166
      Lynn Scarlett

PART III INTERNATIONAL ORGANIZATIONS   189

9   International Organizations and the Improbability of
    a Global Food Regime   191
      Seth B. Thompson

10 Conclusion: The Regional Approach—Reconciling Food Policies
    and Policy Recommendations   207
      David N. Balaam

Index   238

Contributors   245

Tables and Figures

Table

4.1 Soviet Area, Yield, and Production of Grain   50

4.2 Soviet and East European Population and GNP Growth Rates   56

4.3 USSR Grain Exports to Selected Countries: Five-Year Averages 1956–1970, Annual 1971–1977   58

4.4 U.S. Agricultural Exports to USSR by Volume (selected grains): Five-Year Averages 1956–1970, Annual 1971–1979   60

4.5 USSR Grain Imports: Five-Year Averages 1956–1970, Annual 1971–1977   61

4.6 USSR Wheat and Coarse Grain Imports: Five-Year Averages 1956–1970, Annual 1971–1977   63

4.7 Soviet and East European Indices of Total and Per Capita Agricultural Production, 1970–1979   66

4.8 Soviet and East European Division of Labor Force   70

4.9 Volume of U.S. Agricultural Exports, by Selected Commodities, to Eastern Europe   72

6.1 Food Production in East and Southeast Asia   108

6.2 Per Capita Dietary Energy Supplies in Relation to Nutritional Requirements, Selected Developing Countries and Areas   109

6.3 Population and Income Growth   111

6.4   Structure of Population   112
6.5   Distribution of Gross Domestic Product   113
6.6   Annual Changes in Consumer Food Prices   114
6.7   Procurement or Support Prices for Rice in Low-Income
      Market Economies   128
7.1   Food Production Per Capita Indices for Middle Eastern
      Countries   145
7.2   Wheat Yields and Arable Land in Middle Eastern Countries   146
7.3   Gross Domestic Product and Labor Force in Agriculture in
      Middle Eastern Countries   147
10.1 Natural Limitations and Government Intervention Policies   210
10.2 Regional Trade Characteristics   218

Figure 6.1   Estimate of China's Grain Output   121

Acknowledgments

This book began over salsa, chips, and beer at Mayitas in Isla Vista. Our ideas of how to approach food politics evolved over several years. As a result of our decision to use the regional level of analysis, we organized a panel on the topic at the 1979 Western Political Science Association annual meeting. This book contains amended versions of some of those papers as well as others prepared especially for this volume.

We are most appreciative of a number of people who assisted us in this endeavor. Dave Balaam would like to thank his colleagues, especially Paul Heppe and Pris Regan, for their support and encouragement. Michael Carey would like to thank his colleagues David Williams and Seth Thompson for their help. In Tacoma, Nancy Rees, Dan Pearson, and Cindy Connally, and in Los Angeles Mary Hunt and Mary Robinson, helped with research and manuscript preparation. A University of Puget Sound Faculty Enrichment grant covered some of the preparation costs. We would also like to thank the staff of Allanheld, Osmun & Co. for their continuing patience and support. Finally and most importantly, we wish to thank our fellow contributors for their comments and criticism which were freely given. Such an exchange of ideas and camaraderie is stimulating but all too rare.

MJC, Los Angeles
DNB, Tacoma

Introduction: The Political Economy of Food—the Regional Approach

MICHAEL J. CAREY

The politics of regional food policies analyzed here has two goals: (1) to enlarge the scope of study of food problems in the world by utilizing a regional approach, and (2) to make policy recommendations that may not only alleviate human distress but also constitute politically acceptable recourses of action for policymakers. How can the regional approach to world food problems contribute to understanding public policies and proposing solutions to long-term problems? The regional approach collapses previously separate categories of political and economic issues into a more useful category of the political economy of agriculture in various regions. By collapsing these related perspectives, the regional perspective deals with the significant role in food politics played by wealth and power, trade and strategic position, and more broadly, the interplay of public and private power and welfare.[1] In underdeveloped regions, internal differences over the interplay of public and private power are greater than in developed regions. Even so, within the developed regions like the United States, the European Community, and Japan, attitudes toward governmental intervention in the economy differ considerably. The various chapters of this volume point out these differences and similarities and constitute the starting point of the concluding chapter, which evaluates likely outcomes in regional food policies and outlines policy options.

The Political Economy of Agriculture and Food

Food is more than just an economic commodity. Indeed, food has a variety of political dimensions, the most obvious of which is government intervention in food production and distribution. This may take the form of direct intervention, as in the case of nationalized-collectivized agriculture in the Soviet Union, or a more subtle, indirect form, like the manipulation of export credits to favor or hinder the production of certain export products in the Third World. Other examples of more subtle and indirect interventions are research centers, fostered by government grants. Even local governmental zoning or planning for residential development affects future agricultural production in many nations, by influencing what land remains in production.

Another political dimension to food policy is found in trade and aid. Some of the political purposes of national food trade policy are health and safety, taxation, balancing trade deficits, rewarding domestic groups such as urban dwellers or farmers, and maintaining power. A particular vision of world order may be fostered by trade policy—for example, the liberal order fostered by the General Agreement on Trade and Tariffs. Decision makers may also offer food assistance in many forms to other nations to try to foster "good" relations, assist a particular regime, reward or punish, or achieve humanitarian goals. Although the effectiveness of food as a means to attain these ends may be questioned, the political nature of food assistance is clear. Assistance is utilized for many of the same reasons as trade—for example, to reward domestic groups by absorbing surplus crops.

In short, this discussion of the political economy of food leads to the conclusion that there exists a type of "food power" relevant to international relations. As research and development and technology are increasingly being recognized as criticial components of power, it is becoming evident that food is also an important component of political power. This is most obvious in time of war, when the capacity to feed a nation is essential to survival. The chaos produced by recent oil embargoes illustrates that even periods of relative peace demonstrate the attractions of self-sufficiency. The insecurity created by various forms of shortages, including food, indicate that supplies of many necessary goods are less secure and less adequate than formerly assumed during times of relative calm. Even with economic arguments to the contrary, political considerations may dictate the necessity of some form of food security, or at least relative self-sufficiency, to many nations.

As the relative positions of various states change with the waning of the Cold War, and as new issues and forms of wealth emerge, food security provides many with a source of power. Those nation-states with relative food security may be less affected by these shifts in power in international relations. Food security can provide some insulation from various shocks attributed to increased transportation costs, commodity shortages, price rises, intervention in world markets by newcomers, and other supply problems.

Food trade can play a large part in efforts to create a new "climate" in the world when food assistance is used more often for humanitarian purposes. Some types of food assistance have been used to promote agricultural development in malnourished nation-states and at the same time to gain allies. That nation-states, regional actors, or international organizations use food resources to influence international system developments (either through trade or through some form of assistance) also attests to the importance of food and food policies as a means of attaining policy goals. Food however, is not an overt weapon of the type that can be used directly and efficiently to alter behavior in international relations in the manner of military threats, arms sales, cutting off international credits, or destabilizing an economy. Food is a commodity with political implications related to government intervention in production and distribution, and it is used as a tool for aid and trade that contributes to a state or region's political power. Yet we would expect that food can rarely, if ever, be used as a weapon to exhort, punish, assist, pay off, or buy compliance.[2]

Food simply does not lend itself to being used in this fashion for a number of reasons. No one nation or group of nations, or cartel, controls its production and distribution. Too many sellers exist elsewhere in the world for one actor to cut off food to another and expect the "victim" to comply with the "provider's" wishes or demands. Likewise it is impossible for one nation to control prices and supply.

Thus, we would expect that the use of food as a weapon against the oil cartel, or against specific states like the Soviet Union for their incursion into Afghanistan, would prove unsuccessful because these nations are not completely dependent on American grain exports. In addition, rich oil producers could meet price increases in wheat and even counter with discriminatory policies of their own. We would also expect an increase in production and exports of grain by other producers to fill the market vacuum left by exporters who try to withhold food for diplomatic purposes. Also to be considered is the unintended backlash effect by domestic interest groups like consumers, who as taxpayers must absorb higher government support and storage costs, and by producers who stand to lose a "guaranteed" market.

In addition to the nature of markets, many other factors, such as the influence of weather, changes in consumption patterns, and agricultural development in poor regions, can be expected to limit the use of food as a weapon. The availability of alternative supplies is likely to be the most significant limitation to overcome.

A political economy of food, to which, it is hoped, this book will contribute, will shed more light on how to evaluate the use of food as an instrument of foreign policy and demonstrate how to include food in the calculation of political power. Just as international monetary affairs have been more fully understood to reflect political realities and to be both the ends and the means for nation-states and groups of states to influence world order, food should be seen as playing a similar role for nation-states, regional actors, and international organizations.

The Regional Approach to Food Policy

Despite the efforts of international organizations, states, and individuals, issues of food security, hunger, and distribution are no less political today than they were in 1974 when the world focused on these issues at the World Food Conference in Rome. Their grandiose schemes have failed to pass the test of more pressing national concerns.

Food policy at the regional level, largely determined by national and regional actors, holds the key to understanding world food problems. A regional focus also promises additional insight into potential solutions to the political and economic dilemmas of agricultural trade, aid, and malnutrition. The purpose of regional-level analysis is to find a reliable and useful scope for the study of world food problems. Concentration on the world level is simply too general; it lacks focus and ignores significant patterns of behavior that evolve and change at subworld levels of trade and politics, particularly within and between nation-states. By focusing on this too-general level the analyst can fall into the trap of discussing a "world food regime" that has yet to materialize.[3] It is tempting to try to clarify the patterns of trade between nation-states, to chart the influence of GATT guidelines on trade and production, to discover motives behind international organization and national food assistance programs, and to place the whole study under the rubric of a world food regime. That simply produces an overly artificial order out of the chaos of food policies by imposing a semblance of rationality or purposefulness in these myriad acts. Discreet acts by many actors (domestic, national, regional, international, public, or private) do not automatically a regime make. If there is not shared organization, with a minimum common means and goals, then a world food regime does not exist.

One point in favor of the world level or most general level of analysis—which we do not ignore—is that from the world perspective some important trends (and value judgments) emerge, e.g., the fact that enough food is produced in the world to eliminate starvation and malnutrition. Hence, it is imperative to discover what political and economic forces contribute to malnutrition by maldistribution and how the situation may be rectified.

At the other end of the analytical spectrum is the nation-state. To examine world food problems from a national perspective alone would be too narrow. Foreign aid and foreign trade policies are designed to influence actors outside the nation-state. Some nation-states have transcended national boundaries with their production, distribution, and trade policies via regional organizations. A good deal of research has already been done at the domestic nation-state level, especially in the developed countries. While much of that literature is limited to analysis of public policies or interest group–governmental interaction, it is not ignored here; it informs much of work done in this book.

Rather than use a world- or nation-state-level analysis, the focus here is at the regional level. The regional approach, indeed the very notion of a "region" inherent in this level of analysis, is largely a synthesis of political, geographical, economic, and even cultural dimensions that identify certain groupings of nation-states as having something in common—enough in common, in fact, to warrant grouping these nation-states together. The utility of this grouping is demonstrated below. To remind the reader, it should be noted that these dimensions vary in significance when comparing regions. For example, geographical, political, and economic dimensions link Mexico, the United States, and Canada into a region, and the cultural dimension between the United States and Canada is quite relevant as well. In contrast, all four dimensions of a region link the European Community together into a region, in fact, in a much more concrete fashion than in North America. The "regional literature," to the extent that it exists, pays less attention than may be expected to all of these dimensions of a region due to the fact that most regional literature is concerned with integration, and therefore emphasizes the political and economic dimensions that help or hinder the creation of supranational organizations.[4]

It should also be noted that the regional approach here automatically assumes geographical contiguity. In most respects, then, contiguous nation-states or those in relative proximity to each other make up the most "natural" kind of region. Other dimensions fill out one explanation of what a region is. The cultural dimension may include—besides political culture—religion, language, a common colonial heritage, etc. Production and consumption of food is also described as a cultural phenomenon as some regions share consumption patterns. Hence, even in regard to varieties of food some basis for a region may be identified.

Other reasons exist that point to the utility of this approach. One is similarities in climate, production, etc., and in developmental patterns or potentialities. Still another justification relies on the area-studies rationale, even though that may be one largely of convenience.

Trade and political relationships clearly exist at the regional level, e.g., in the European Community, between ASEAN members, and between the Soviet Union and Eastern Europe. The European Community has moved farthest in institutionalizing regional patterns, but as the chapters here demonstrate, food policy has a strong regional dimension in many areas such as North America, Latin America, the Middle East, Asia, and sub-Saharan Africa.

Who are the main actors at the regional level? Nation-states and regional organizations constitute the main actors in food policy. Generally, the dominant actors are nation-states, just as in most other international relations, who, in regard to food policy, attempt to act independently and/or in coalitions or other forms of cooperative behavior. As will be seen, neither conflict nor cooperation dominates the food arena. Interest groups and farmers are key actors influencing both policy formation and implementation; more often than not their main arena is within the nation-state. One important exception is found in Western Europe in the European Community. Internationally, distribution is often effected by private companies, which usually have a national headquarters. Truly international actors such as agencies of the United Nations are rather limited in effect and even purpose, as is explained in the chapter below on international organization.

Conflict or cooperation may be the order of the day, due to the nature of the products, efforts, and influence of interest groups, parties in power, and intervening political purposes and interests (e.g., strategic). Within some regions, e.g., the European Community, cooperation tends to be dominant, as mechanisms for conflict resolution exist, are effective, and are used quite often. Interregional trade tends to display more conflictual behavior, e.g., in East-West or North-South trade, but at this level mechanisms to create the rules of the game do exist in GATT or UNCTAD conferences.

Which are the important regions of the world that have evolved and continue to evolve complex regional patterns of trade and aid interrelationships? The relatively developed countries, often referred to as the North (excluding the Soviet Union, Eastern Europe), constitute the main highly interdependent areas of the world: Europe, North America, Japan (and to a lesser extent Australia and New Zealand). This is the main trading area of the world and features very general political-economic similarities. These actors often cooperate or coordinate approaches when dealing with other regions in the underdeveloped South or with the Communist states.

These Northern independent regions share many common interests and political values and have strongly influenced the nature of world trade, especially in agricultural products, even though they are divided into various groupings characterized by internal and foreign policy differences that preclude a monolithic trade block and inhibit its emergence as an effective world power.

The Soviet Union and its allies constitute a region that intervenes in world food markets in fits and starts; if all rich nation-states have an "obligation" to assist poor nation-states, they should be involved in any analysis of regional aid. The Eastern European region is now important for its potential as a market, producer, and, in the future, as a source of food assistance to poor regions.

The poor states of the South, whether taken as a whole or divided into specific regions (usually by continents) defy easy categorization and give rise to a set of serious questions for the analyst. These nation-states often trade more with the North than with their neighbors, frequently selling their luxury food products abroad, while importing basic foodstuffs, and at times struggling to feed themselves while making efforts at internal reform. These states also make many demands in international forums for reform of the world trade system. States in the Southern regions differ dramatically in levels of ability to feed their populations, involvement in world trade, attitudes toward investment in agriculture, and the nature of political and economic relationships with nations outside their region. Since poor regions are faced by a plethora of problems many of which are only now being clearly identified, dependency models and other "solutions" are nearly irrelevant today because they are premature and/or simplistic. To some extent, an awareness of the *magnitude* of agricultural differences and the *number* of potential solutions via public policies are the only common elements of a regional food policy in the poorer regions.

Given this background, the following chapters explore these topics in a variety of regions. Robert Dickens and Richard Moore examine the prospects for better coordination of North American food policies. They argue that there are certain advantages to a better managed U.S.-Canadian-Mexican policy given structural elements of national agricultural conditions. They believe that "regional coordination and management is likely to occur because it is both politically desirable and economically beneficial." At the same time, certain issues and differences prevent the realization of a truly regional policy. This chapter underscores the importance of agriculture policies to the United States, Canada, and Mexico; it also emphasizes the extent to which food policies, especially in the area of trade, must be successfully managed.

Michael Carey assesses the "rules of the game" formulated by the European Community to deal with agricultural problems. Decreasing bi-polarity and increasing interdependence have put the institutionalization of a European regional food policy on a roller coaster not unlike that which has resulted from attempts to achieve political integration. The discussion of European multilateral and national bilateral relations focuses on many of the internal dilemmas that the community faces. This chapter, along with the North American chapter, highlights a number of controversies surrounding agricultural politics in advanced democratic nations that are marked by not only wealth and welfare problems but strategic issues as well.

Balaam and Carey discuss the nature of Soviet and Eastern European food policies, which diverge from the basic collectivist model once posited by the Soviets. Reliance on the West for supplies contradicts the basic drive for autarky in the region, a position that may be "softened" enough in the future to allow this region a permanent place in agri-trade in the world.

Richard Ganzel discusses food policies in Latin America, focusing on Chile, Brazil, and Argentina, where government food self-sufficiency poli-cies have not necessarily increased food production or distribution to those who need it most. He believes Latin American countries would benefit from policies that exploited comparative advantages: "the point remains that dedication to efficiency, not to self-sufficiency for a predominately agrarian society, must be the goal of those who would help the urban and rural poor."

David Balaam, on the other hand, underscores the role of food in development goals in Asia, a region marked by diverse national political, economic, and social systems. This chapter focuses on agricultural policy arenas in Japan and China and on conditions in a number of Southeast Asian nations. A regional food policy is beginning to emerge in Asia but may be deterred because domestic and external constraints pressure governments to pursue traditional development strategies or because regional security still concerns policymakers.

Marvin Weinbaum explores agricultural constraints in the Middle East, where few if any countries are destined to become self-reliant in meeting their domestic food requirements. The impact of political rivalries and bureaucratic policy in this region makes regional cooperation improbable in the near future.

Lynn Scarlett analyzes the impact of a variety of food policies in tropical Africa, a region with some of the most critical supply problems to overcome. The scope of policies within these nations, regionally, and vis-à-vis the

North is of special significance to local solutions to the region's agricultural problems.

Seth Thompson's chapter discusses the role that international organizations, particularly the FAO, play in the global food system. A food "regime" is not becoming institutionalized. Not hopeful that IOs will be able to dent the food problem, he sees their purposes as providing "an established, focused, and continuing arena within which the politics of food can be played out."

Finally, the concluding chapter analyzes the prospects for cooperation in the face of vastly differing policies in the various regions. We believe the likely outcome is continued if not intensified irreconcilability of policies with few exceptions, contingent on an ever-increasing nationalistic focus that has spread to many of the developing regions and due to pressing wealth and welfare problems in the developed regions. Security concerns cannot be overlooked, and their reemergence near the top of some national agendas postpones efforts to solve hunger problems. Policy recommendations must be made in this light since policy decisions, at any level, are first of all political. That is, the politics of food from a regional perspective highlights many of the trade-offs decision makers will have to choose between, given the ends they seek to accomplish, and the means they have at their disposal to accomplish those goals. Policy recommendations cannot be divorced from the political environment in which they exist.

Notes

1. Key examples of the literature on the analysis of political economy are: Jacob Viner, "Power versus Plenty as Objectives of Foreign Policy in the Seventeenth and Eighteenth Centuries," *World Politics*, vol. 1, 1 October 1948; Richard N. Cooper, *The Economics of Interdependence* (New York: McGraw-Hill, 1968); Robert Keohane and Joseph Nye, *Power and Interdependence* (Boston: Little, Brown and Co., 1977); Klaus Knorr, *Power and Wealth* (New York: Basic Books, 1973); Edward Morse, *Foreign Policy and Interdependence in Gaullist France* (Princeton: Princeton University Press, 1973), pp. xi–105.

2. See for example "Food for Crude" in *Today*, 27 July 1979, p. 9.

3. For a discussion of the "regime" aspect of food regime see Raymond Hopkins and Donald Puchala, "Perspectives on the International Relations of Food," *International Organization*, vol. 32, no. 3 (Summer 1978).

4. Two very penetrating analyses of integration-regionalism are found in Reginald J. Harrison, *Europe in Question* (New York: New York University Press, 1974), and Charles Pentland, *International Theory and European/Integration* (New York: Free Press, 1973). For analyses that include culture see Amitai Etzioni, "Political Culture and Integration in Scandinavia," and Werner Levi, "Political Culture and Integration in Southeast Asia," in Paul Tharp, ed., *Regional International Organizations/Structures and Functions* (New York: St. Martin's Press, 1971).

Part I
The Developed Regions

2

Food Policy in North America: The Bread Basket

ROBERT E. DICKENS AND RICHARD K. MOORE

The purpose of this paper is to examine the reality and possibilities of North American (Canadian, Mexican, and American) food policy. We will explore the prospects for increased "rationalization" of existing policies. Our concern is the political economy through which man provides himself with food and fiber. Although form and direction of food policy are shaped by political and economic concerns, the affective and cognitive tenor of food policy is heavily moderated by social and cultural factors. Our analysis will attempt to weave these strands together in both descriptive and, finally, prescriptive fashion.

There are several reasons to support the notion that these three nations may be regarded as an agro-economic region. Geographical continuity, soil, climate, pests, and water argue in support of a North American regional perspective. Economic relations among the three nations reflect interdependence as well as competition. Historically, both Canada and Mexico have had strong commercial links to the United States. U.S. investments dominate both countries. The United States is the major source of foreign currency earnings for both. Between Canada and the United States, the "culture" of agriculture is similar in both structure and attitude. Between Mexico and the United States, the links are more purely economic than cultural. Migrant farm labor allows American enterprise a profitability it would not otherwise enjoy, while simultaneously relieving unemployment pressures and providing foreign earnings in and for Mexico.

Despite some rather obvious and important links among the three nation-states under consideration, significant differences exist in the nature, purpose, and structure of agricultural policy.

It is widely accepted that agricultural issues consistently play a central role in domestic policy-making arenas. Cycles of varying intense political conflict have accompanied the establishment and periodic adjustment in agricultural policy. Of particular importance have been links between agricultural policies and their associated national and international markets. The sense of urgency in many writings on this topic conveys sentiments long familiar to the history of agriculture in nation-states, yet amplified by a newfound "global" perspective. The world importance of agricultural issues has intensified as a result of population pressures, desertification, drought, war, famine, starvation, and land hunger.[1]

Increasingly, however domestic agricultural policy is directly or indirectly linked to foreign policy.[2] This not only includes traditional areas such as terms of trade, but has also recently involved strategic considerations with regard to the superpowers, China, and the Middle East.

The Regional Perspective

A regional approach to the agricultural policies of Canada, Mexico, and the United States necessarily assesses trilateral relations between nations that are both competitors and long-standing trading partners. Furthermore, such an approach compels scrutiny of political institutions, and public policies in a setting illuminated by production and productivity, the structure of the agricultural sectors, and the role and purpose of national government must be examined in each national system as a basis for understanding the potential of a North American region.

Given the changing context of agricultural policy, certain valid questions arise. Will the dynamics of the energy situation give rise to an essay at a countervailing food producers' cartel? Do the implications of an emergent North American food policy extend to other resources and issue areas? Or vice versa? Strategically, can the three nations, acting in concert, penetrate European markets that all have identified as unnecessarily restrictive? How does regionalism "fit" long-term, bilateral relations (e.g., Canada-China, United States–Japan, Cuba-Mexico) with states beyond North America? As major grain-exporting nations, can the United States and Canada effect equitable trade policy with petroleum-rich Mexico, whose policies have identified it with the calorie-deficient Third World? Finally, can the powerful distraction of nationalism be overcome in the quest for a regionally rationalized food policy?

Agricultural Policy Goals

In both Canada and the United States, agricultural policy goals try to provide farm income maintenance and price stability. In Mexico, on the other hand, no such simple formula may be accurately enunciated. Two agricultural systems exist concurrently in Mexico. In southern and central Mexico a system of communal (*ejidal*) agriculture exists. Land use and control under the *ejidal* system is a major social and cultural goal in and of itself. In the north, the great estates *(haciendas)* tend to predominate. These large enterprises have proven to be productive and to be amenable to mechanization and modern irrigation; they are an important source of foreign earnings. Mexican agricultural policy, in consequence, is bifurcated, with somewhat contradictory social and economic goals.

Historically, agricultural development in Canada and the United States has been characterized by (1) homesteads as a lure to European immigration; (2) aid to foment rural social and economic development; (3) engaging in agricultural research and dissemination of findings through extension services and experimental farms; and (4) targeting agricultural policy, in general, on the family farm as the basic production unit.

If, in the United States, and to a lesser extent in Canada, the agricultural sector provided what Horace Greeley labeled a "safety valve,"[3] then the Mexican agricultural sector can be viewed as something of a "burning fuse." Land hunger has led to fragmentation of the great land holdings of the ancient creole ruling families. The great parcels, often iniquitously gained and exploited, nevertheless had the often realized potential of competitive productivity on the world market. As Mexico's limited amount of arable land is divided and subdivided to satisfy pressing social and political demands, the ability of modern agricultural techniques to satisfy economic demands is diminished. Land hunger in a booming populace remains, populations must be fed, and soil resources are finite.

In all three nations populations were dispersed geographically because of agriculture. In Canada and the United States these populations would probably otherwise have been clustered near the seats of political and commercial power and have hindered stable political development. In Mexico, a Spanish and creole ruling class forcibly expelled a native American ruling class and converted communally held agricultural lands into private ownership. In the former cases, central agriculture policies contributed to stable development whereas, in the latter, land use has been perhaps the major cause of domestic instability.

Agrarian protest in Canada and the United States has been real and sometimes violent. Its counterpart in Mexico has been frequent and almost always violent. In the northern tiers, protest has involved landowners seeking amelioration or subsidy. In the southern tier, protest has involved the landless seeking land and social justice.

In the United States, for example, the institutionalization of conflict relationships between agriculture and government during the New Deal assured farmers a say in policymaking.[4] In Canada, however, the picture is not as clear. Fowke suggests that historically Canadian agriculture has had weak links to policymaking arenas.[5] Theoretically, at least, the linkages between the Mexican peasant farmer and his government are direct and decisive. The reality is less sanguine. Half of the cultivable land remains in the hands of a tiny fraction of the population. The other half is made up of numerous small enterprises that lack credit, technology, and infrastructural support.

In Canada governmental-agricultural linkages are the result of farmer initiatives. Nevertheless, the character and effectiveness of agrarian-sector access to key decision makers remains an open question. A recent Organization of Economic Cooperation and Development (OECD) study suggests that recent changes in certain Canadian agricultural programs were designed to help the agricultural sector adjust to a "variety of destabilizing influences emanating both from elements in the domestic non-farm economy, from the international economy and from natural causes. Since 1973, Canada and its major trading partners have experienced periods of rapid economic transition."[6] Recovery from economic slowdown and recession continues to be vulnerable to external factors like oil prices and threats to supply, such as that seen in the Iranian situation. Nonetheless, adjustments in Canadian agricultural policy indicate the power of a farm bloc and the responsiveness of the Canadian parliament.

In Mexico, several governmental agencies and a major sector of the ruling political party (*Partido Revolucionario Institucional*—PRI) are designed to receive and translate inputs from the agricultural sector, yet these mechanisms fall short of adequacy.

Three agricultural variables and one nonagricultural variable impinge heavily on the Mexican political economy. First, the government must find ways of slaking the continuing land hunger of the *campesinos*. Second, the foreign exchange earnings from the sale of capital-intensive agricultural produce must be protected and increased. Third, policies must be developed that will make Mexico independent with regard to supplying its domestic market for staples (namely, corn). The final and nonagricultural variable is the rich petroleum potential, which gives policymakers some latitude for discretion in dealing with the preceding three variables.

GLOBAL DEPENDENCE

North American agricultural policies, especially in Canada and the United States, have additional similarities with respect to world food supplies. Both share the blessing and burden of large, dependable agricultural surpluses in a food short world. As Lester Brown reports:

Before World War II, both Latin America and North America were major exporters of grain. During the late 1930's, net grain exports from Latin America were substantially above those of North America. Since then, however, the failure of most Latin American governments to make family planning services available and to reform and modernize agriculture have eliminated the net export surplus. With few exceptions, Latin American countries are now food importers. Over the past three decades, North America—particularly the United States, which accounts for three fourths of the continent's grain exports—has emerged as the world's bread basket. Exports of Australia, the only other net exporter of importance, are only a fraction of North America's. The United States and Canada today control a larger share of the world's exportable supplies of grain than the Middle East does of oil.[7]

To this must be added the fact that Mexico, with its large petroleum resources, could become self-sufficient in staple foods in a relatively short time.[8] Taken as an export-producing region, North America must be regarded as the most powerful economic entity of its sort. The effectiveness of that power, however, is predicated on the extent of unanimity or harmony of the several, respective foreign economic policies.

Here, too, Canada and the United States resemble one another, while Mexico stands apart. In the United States, for example, "since the early 1970s most farm commodities have moved in the free market from price support operations. Farm price programs have been becoming more market oriented."[9] Likewise,

throughout the period under review, 1973/77, Canadian policy makers retained the basically market oriented approach to agriculture that prevailed during the late 1960s and the early 1970s. It is clear that Canadian officials feel that price costs signals should reflect the market on a world basis, rather than on a strictly interior. For most products, prices recieved have been at a level that cleared the markets for these commodities.[10]

However, Mexican foreign trade policy, in general, emanates from a protectionist, developmental philosophy. As in most things political, Mexico's foreign economic policies are dominated by the will of the president of the country. The desire for self-sufficiency, if not autarky, seems to have a persistence that transcends individual administrations despite the power of the president to dominate in a given instance. In consequence, the world market for grain may influence Mexican agri-trade policy but has not been the sole determinant of that policy. Mexico is unlike Canada or the United States in that political, rather than economic, factors seem central to the development of food strategies.

AGRICULTURAL SUPPORT: ORIENTATION AND MECHANISMS

Canada and the United States rely on global food dependence to maintain domestic agricultural prosperity. Only a portion of the Mexican agricultural sector has such a reliance. The remainder either is subsistence or exclusively serves the domestic market. Canada and the United States have adopted policies that display something of a commitment to open international markets. Economically, Mexico must pursue policies that will achieve agricultural independence and earn foreign revenue. In a sociocultural sense, Mexico must pursue policies that at least appear to slake the land hunger of millions of peasants. Politically, policies must be pursued that seem to move the Republic away from its "suffocating" dependence on the economy of the United States.

Although Mexico is clearly distinguished from its northern neighbors, one should not be confused by the belief that United States and Canadian agricultural goals are identical. Canada and the United States may have similar values concerning the desirability of market economies, but they nonetheless define the role of government in that market in substantively distinguishable terms. Canada has made adjustments in its Agricultural Stabilization Act (1975) and its Western Grain Stabilization Plan (1976). "In all major programmes, focus has been shifted away from strict market price stability, towards stability of the margin between revenue and total cash costs. Under the auspices of these major programmes [ASA and WGSP] it is the difference between prices and cash costs which are being stabilized."[11]

The United States, on the other hand, has maintained a commitment to interventions keyed to target prices. Since international agricultural market prices for most commodities have been above the target price level in recent years, deficiency payments have declined, thus reducing the cost of price support programs from "a level of almost four billion dollars in 1972 to around a half billion dollars annually in 1974–1976."[12] Increasing inflation and farm input costs may not be reflected in prices determined in international food markets, which are subject to production uncertainties and bumper harvests, both of which lead to prices below production costs. U.S. farmers find their production costs increasing because of both internal and external factors. These increases are not always reflected in the income support strategies of U.S. agricultural policy, whereas Canadian agricultural policies are more sensitive and responsive to their farmers' cost-price squeeze than their U.S. counterparts.

As both a developing nation and an inextricable member of the global and regional food community, Mexico is characterized by farmer-government relationships that are more byzantine than those of her northern neighbors. To a great extent, world market forces control pricing and productivity in the agricultural sector. The major exception lies in the domestic supply of basic foodstuffs. A national supply company (*Compañía Nacional de Susisstencias Populares*—CONASPUP) ensures the availability of an adequate supply of food for consumers at an affordable price. This role occasionally involves CONASUPO's going abroad for needs if there is a domestic shortfall such as that which occurred in the 1979–80 crop year. Direct governmental intervention in production, rather than the marketing sector, has been the chosen vehicle of Mexican regimes for the past five decades. Loans, credits, technical assistance, irrigation projects, and land redistribution are the generally preferred policy outputs. These outputs have been intended to provide either domestic self-sufficiency or foreign exchange earnings, not to protect the individual producers from the vagaries of the international market.

PROGRAM PARAMETERS AND SIMILARITIES

Parallels in Canadian and U.S. agricultural policy extend beyond orientations toward international food markets. In the area of consumer protection and satisfaction, northern North America is beginning to develop some crude commonalities with Mexican consumer food prices.

Canadian price support programs have recently undergone reexamination due to a food surplus crisis in the grain-producing prairie region and livestock market instabilities. According to Warley:

The broad result has been an explicit reaffirmation and intensification of previously implicit policy. It is now the avowed intent of the federal government to foster a commercial agricultural industry that is market and development oriented, internationally competitive, and economically viable without sustained transfer payments.

Government has, however, accepted an enhanced role in indicative planning, in facilitating the product mix of agriculture, in promoting market development and diversification, in fostering supply management, and providing enhanced stability.[13]

These policy goals are being effectively achieved through a series of adjustments, which the OECD refers to as "generally good compromises." "Considering consumer, farmer, and national trade interests that would need to be balanced in a period of basic instability, the attempts to stabilize margins for agriculture should be seen as workable."[14]

Amendments to the Agricultural Stabilization Act (1975) and the Western Grain Stabilization Plan are cases in point. As mentioned above, the purpose of these acts was to shift away from merely stabilizing prices toward stabilizing the margins between cash costs and returns. This adjustment was to be accomplished through deficiency payments based on 90 rather than 80 percent of listed commodity prices averaged over the previous five, rather than ten, years. Shortening the base period and increasing the market price percentage "appeared to allow basic resource costs and efficiency to determine the longer term structural development of the sector, while ensuring that this adjustment would take place within a flow of net cash income, stable enough to be neutral with respect to the (structural) adjustment process."[16] Further, the Canadian federal government entered into responsibility-sharing agreements with the agricultural provinces affected by these measures. Support levels could be increased above the levels provided by the 1975 Act under three conditions: increased costs for provinces/producers; benefit equity among producers; and understanding that supports should not spur overproduction. The OECD study finds that:

the means for designating specific crops on an intermittent basis (fruits and vegetables) and offering deficiency payments to producers of selected commodities, can give significant protection to the short-term income situation facing small segments of agriculture when needed, while maintaining very low levels of government transfers to the agricultural industry as a whole.[16]

The Western Grain Stabilization Plan of 1976 was similar to ASA in many respects; however, the funding mechanism differs specifically. On a sector basis, producers pay a small portion of their cash receipts, which are augmented by government contributions into a Stabilization Account. If necessary, and as of 1977 it had not been, the fund would make deficiency payments to participating producers from the grain and oilseed sector up to an amount that would be sufficient to raise total sectoral net cash flow to a level equal to the average of the preceding five years. The total amount is divided among participants in proportion to their average contributions in the year of the payment, plus that of the previous two years.

Price stabilization has also been a component of U.S. agricultural policy. This is seen in the various support levels with which policymakers have experimented since the first Agricultural Adjustment Act (AAA) of 1933. The goal of this policy, like that of Canada, has been to maintain farmer purchasing power or real income. The parity system established a 1909–14 base period upon which price supports would be constructed. The range of allowable parity levels fluctuated from 52 percent in 1936 under the second AAA to 90 percent under the Agricultural Act of 1954. Parity and its contemporary meaning remains a heated issue in agricultural policy arenas.[17] In addition to price supports, other techniques have been used. Supply management was tried during the Kennedy administration but was unsuccessful because of farmer resistance to production controls. The Food and Agriculture Act of 1965 was a departure from the stalemates of the 1950s over the proper mix of price supports and "free markets." The 1965 Act authorized price supports close to world levels, thus more closely linking domestic with international food markets. Income supplements were provided to farmers who voluntarily reduced their crop acreage. The result was increased farm income and reduced government-held surpluses. The Agriculture and Consumer Protection Act of 1973 substituted target prices for conventional price supports. This mechanism is generally agreed upon, though controversy continues as to the proper level of target prices and their apparent lack of linkage to production costs.

In a sense, Canadian policies aimed at the cost/price margin seem more refined and clearly directed toward the purpose of increasing farm income in a manner that does not threaten further gains in agricultural efficiency and rationalization, through structural realignments caused by the withdrawal from agriculture of marginal farms. United States policies, in contrast, are imperfectly tuned to externalities affecting farm incomes such as inflation, capital costs, increasing input costs, and market uncertainties.[18]

Barely noticeable in the discussion of changing U.S. agricultural support policies is the inclusion of a new element in the 1973 statute, "consumer protection." This change reflects the dynamic of domestic political power in the United States. Reduced numbers of producers and increased numbers of self-conscious consumers added a major new variable to the agricultural policymaking process. Urban consumers have become a force to be reckoned with, and both "lower ceilings" and "higher floors" for food prices are the effective, contradictory result. Consumer political strength seems destined to grow at the expense of producers in the future.

In Canada, the consumer "movement" has not achieved either the power or the success of its neighbor. Nevertheless, the pattern of urbanization, consolidation of small producers, automation of agricultural methods, and rural-to-urban power shifts suggest that this factor will become increasingly important.[19]

Mexico's political sensitivity to its burgeoning urban population, its subsidy of consumer food prices, and its general insensitivity to farm price supports allows the inference that there will be an increased commonality of effective domestic food policies among the three North American states, albeit for wholly different reasons.

One further commonality in regional agricultural issues is the problem of labor. The U.S. dependence upon thousands of Mexican laborers is well documented.[20] In Canada, as well, agricultural labor supply has been perceived as one of three major limits to increased production.[21] Logically, a rationalization of the labor relationships between the three nation-states would provide something of a basis for intraregional agricultural policy, although such an understanding does not appear imminent.

There are broad similarities between the U.S. and Canadian farm price support programs, although significant differences in detail remain. The rise of consumerism in the United States is having the effect of putting the United States' price support program in a direction roughly coincidental with that of Mexico. Mexico's protection of producers' profitability is virtually nonexistent. Mexico is a source of virtually unregulated agricultural labor for the United States and could become so for Canadian agribusiness if political agreement could be reached or dire circumstances prevailed in either nation.

RESERVE POLICY

The role of government in regard to agricultural reserves provides additional similarities between the United States and Canada, while Mexican views stand apart. The United States has long been not only the world's major exporting nation but also the world's major granary. The Commodity Credit Corporation, for example, continues to guarantee the existence of a market for farm commodities complete with publicly owned storage facilities for excesses that cannot, for political and economic reasons, be absorbed elsewhere.[22] There is a good deal of controversy on this point. New Deal liberalism[23] suggests that government purchases and storage programs removed some uncertainty from the farm sector, stabilized prices, and ensured supplies. To others, such a policy constituted a political intervention in market affairs, thus skewing outcomes with growth-controlling inefficiencies, suppressed production, and neglected cultivation of international markets and trade.[24]

A reconciliation of these viewpoints seems distant. Such questions of political economy are likely to be answered in normative terms. What remains a fact is that, in the United States, government-held reserves are tangible evidence of our agricultural productivity. The existence of that reserve (surplus), it seems on balance, has constituted a source upon which recent demands could be satisfied. Perhaps that reserve also provides time for subsequent adjustments to erratic and extreme market fluctuations.

In the United States, the relationship between private farmers and government, as seen in the reserve policy, is marked by symbiosis. Such policy has been legitimate and carefully formulated to serve subsystemic constitutents.[25] The rise of the American consumer movement may upset this fragile tranquillity in the future, but it has not done so as yet. Government-held reserves could serve many interests to the detriment of others.

With regard to Canada, the situation is historically and structurally different. In Canada, commodity reserves are farmer-held. Storage and transportation costs are born by the producers. Release of reserves is regulated by quotas mandated by the Canadian Wheat Board, thus making storage and transportation of reserves a direct cost of production. In the words of one Canadian farmer, "we get no storage payments or loans or any other help, I'll trade places with (U.S.) growers anytime."[26]

Mexican reserve policy is as elusive as its surplus. Despite Mexico's status as a leader, in terms of economic development, of the Third World (the Republic's GDP is ahead of those of Greece and Belgium and just behind that of Australia), it has chosen to become something of champion of the LDCs rather than a novice member of the "haves" of the Northern Hemisphere. During his tenure (1971–76), Mexican president Echeverria formulated and espoused a "Charter of Economic Rights and Duties of Nations." This proposal called for a rather greater "North" responsibility to the "South" than simple maintenance of an international grain reserve, although the latter would have been encompassed by the proposal. Since his accession to office, the current president, Lopez Portillo, has continued a somewhat more muted advocacy for the rights of the Third World, without specifically championing an international grain or food reserve. On the other hand, both the current administration and most sectors of Mexican society, ever jealous of the outward signs of independence, have rejected calls for a North American Common Market.[27] Presumably, this rejection would include, for the moment, entry into a North American food consortium.

Regional Interdependence

Another reason for considering the agricultural policies of Canada, the United States, and Mexico in a regional perspective is the nature of the relationship among these nations. Keohane and Nye find that "Canadian-American relations fit closely the conditions of complex interdependence."[28] In terms of the characteristics of complex interdependence established by these authors, Mexico and the United States also share this relationship.

The absence of military force is important. Violence between Canada and the United States ceased about 1895. That between Mexico and the United States effectively ended in 1917. The period thereafter has been marked by both close wartime alliances and some efforts by Mexico and Canada to maintain a measure of independence from the United States with respect to national security affairs. Canada, although an active member of NATO and an active participant in the North American Air Defense Command, has participated in virtually every United Nations peacekeeping operation. Mexico, although generally neutral in most Cold War matters, was an active belligerent during World War II, opposing both German and Japanese forces.

The importance of the trade relationship among the three nations cannot be understated. Whereas Canada is unrivaled as the principal trading partner of the United States in both imports and exports, Mexico's trade relations with the United States are not insignificant, especially from the Mexican perspective. (Mexico ranks tenth as a supplier to the United States and fifth as a market). Most of Mexico's and Canada's trade is with the United States.

Indeed, relations between Canada and the United States are sufficiently close for Keohane and Nye to have tested it against the notion that economic processes would ultimately lead to a degree of regime changed marked by increased political integration.

Certainly economic integration was increasing. Exports to the United States rose from half to over two-thirds of Canada's total exports between 1948 and 1970. Each country was the other's largest trading partner, and their exports to each other rose from 26 percent to 36 percent of their total exports during the 1960's.[29]

Political integration, however, has not occurred due to nationalist trends in Canada and the growing infrequency of successfully linking economic and trade issues to other bilateral concerns. "Both governments were aware of the welfare losses they would incur from a disruption of economic integration, and of the necessity of some policy integration—preferably informal—to maintain the economic system."[30]

Much the same thing can be stated of the economic relations and political permutations between the United States and Mexico. Although the economic connection between the two has been rather onesided in favor of the United States, the connection is real and vital both ways. Labor, vegetables, coffee, petroleum, natural gas, and ferrous and nonferrous metals flow north. Tourism, investment, industrial goods, and technology flow south. Political relations, if not cordial, have usually been "correct." Border and parliamentary commissions between the governments exist at the general policymaking level, while specific, functional areas between policy implementers are cordially and mutually beneficial. Examples of the latter include desalinization, health, animal inoculations, insecticides, law enforcement, and penology.

Even more than Canada, Mexico has been reluctant to move toward bilateral integration with the United States in political sectors because of fears of losing any measure of sovereignty to the "colossus." Informal discussions concerning the formation of some sort of a regional "common market" seem to have ground to a halt with the growing Mexican belief that closer integration "whatever its effects on the growth of Mexican industry and employment, would result in the total subjection of the Mexican economy to U.S. control."[31]

The tenor of bargaining between the United States and its neighbors has matured from one characterized by attempts of the United States to exert its preponderance to its opposites' learning how to use growing nationalism and public politicization (of issues) to achieve greater gains than they would have in past times. Relationships have eschewed formal "linkage" because such a formal connection is simply insuperable in the domestic political arena.

On the part of U.S.-Canadian relations, what has emerged in the 1970s is a stable regime based on "awareness of the potential joint losses from disrupting economic interdependence and acceptance of the important role of informal transgovernmental networks in managing relations under conditions of complex interdependence."[32] Rather than integration, what has occurred between Canada and the United States is a marked degree of intergovernmental coordination conducted by ongoing relations between federal agencies, as well as some states and provinces. "The classic image of governments interacting through foreign offices is clearly inappropriate in the Canadian-American case."[33]

A similar statement could be applied to the case of U.S.-Mexican relations; however, it would belie the deep distrust and hostility toward the United States held by a significant portion of Mexico's leading citizens. Too many generations of disdain and inequity have passed for these feelings to be quickly shed. Octavio Paz has stated it succinctly.

As for its relationship with the United States, that is still the old relationship of strong and weak, oscillating from indifference and abuse, deceit and cynicism. Most Mexicans hold the justifiable conviction that the treatment received by their country is unfair.[34]

Perhaps time and the recognition of an agricultural/energy complementarity will end the sense of asymmetry, but such changes will not come soon and will require both tact and candor on both sides.

Complex interdependence characterizes the agricultural sectors of all three nation-states. Canada and the United States share a certain historical continuity of agricultural policy. Writing about Canada, MacFarlane and Fischer[35] indicate that emphasis in farm policy was placed on education, extension, and research conducted by the public sector; public provision of farm credit aimed at the family farm; grading and inspection services; public assistance in the provision of farmer-owned warehousing and processing; price supports; crop insurance; assistance to the rural poor; the establishment of national marketing entities; and the maintenance of close ties to the farm movements. Practically all of the above are found in the U.S. agricultural policy, the key difference being the absence of a wheat marketing board, although one might argue that the Commodity Credit Corporation, the Office of the General Sales Manager, and/or control of export supplies by U.S.-based MNCs leads to similar market distortions.[36] Although the priorities are radically different, Mexico's agricultural policies cover most of the functional areas previously mentioned. The major functional differences are in the area of foreign marketing, the equitable distribution of government services (the north fares far better than the south of Mexico), and strong orientation to consumer welfare rather than to the farmers.

All the nations' farm sectors display a tendency toward larger, fewer, and more specialized farms as a result of the impact of technology, resource mobility, financing, and risk management.[37] In the case of Mexico, however, the above economic fact exists in direct contradiction of governmental, political, and constitutional dictates. Breaking up large estates and land redistribution is supposed to have been the norm.

For Canada and the United States, the result has been that problems such as overproduction, declining relative farm income, and price supports have been increasingly linked or turned over to the market to permit and promote rapid structural adjustments. Mexican policy, by way of contrast, sees the government intervening to control consumer prices and remaining aloof from the foreign market with regard to exports. Consequently, discussions of free trade zones for agricultural products between Canada and the United States have become increasingly frequent,[38] while those between Mexico and the United States seem to have led to something of an impasse.[39]

TRADE LIBERALIZATION

Perhaps no other issue in the agricultural policy arena is more clouded with rhetoric and convoluted meaning than that of "free trade." Shuman takes the view that the solution to world hunger lies with free-market solutions, which have been distorted by governmental intervention.[40] In this view, national and international politics serve as disincentives to the efficient production and distribution of agricultural products, the ultimate losers being hungry LDCs. A less extreme view is held by Seevers, who contends that governmental regulations separate individual national markets, but that, particularly with regard to grain, there is an internationally available price reference mechanism embodied in futures trading that approximates a global food market.[41] There are, with regard to this issue, certain givens. Largely because of domestic restraints, the free-trade solution has seldom made its way undiluted into agricultural marketing policy. Examples include Europe's CAP, as well as the aforementioned programs of the United States and Canada.[42] What is clear from these examples is that international food policies are constrained by domestic political parties, opinion, and interest groups, all of which desire to protect a domestic endeavor that addresses the most basic element in the hierarchy of human needs. It is also evident that agricultural interests and government policymakers are characterized by a high degree of interrelatedness.

While that close relationship has undoubtedly led to distortions in ideal-typical market relations, it is equally true that, over the long run, the very closeness of these systems had major positive effects. In Canada and the United States, it has promoted the development of the world's most productive (both relatively and absolutely) agricultural sectors, while in Mexico it has provided a political stability that, if not just, makes the reality of self-sufficiency and economic development a short-term possibility. Results of this relationship have in many instances been inequitable; food has been produced in sufficient quantities to make it a major source of foreign trade earnings and, for Canada and the United States, a major resource in international relations. Further, it is also clear that agriculture defies rationalization. One need only look to the reality of events for evidence. Market uncertainties involve such things as climatic variations, political embargoes, change in agricultural orientations in established commodity markets, and cycles of protectionism.

It would appear that it is the character of trade liberalization, rather than trade liberalization itself, that is of the essence. None are likely to move toward policies recognizable as complete rationalization of agricultural production and distribution. Instead, the terms under which respective comparative advantage can be increased or maintained will mark agricultural policy adjustments. Recent Canadian-U.S. policy adjustments indicate

ALBRIGHT COLLEGE LIBRARY 183819

incremental moves toward a closer linkage between market forces and agricultural policy. This has been accomplished, however, in an atmosphere of climbing demand and progressive structural adjustments in their respective agricultural sectors.[43] The drought-caused failure of the 1979–80 Mexican corn crop evoked massive governmental entry into the world market just as the United States was imposing its feed/grain embargo on the U.S.S.R. Both Mexican and U.S. actions in this instance show the sensitivity of political factors to a host of agro-economic variables, in quite unexpected fashion.

Domestic stability has been a prerequisite for productive change as regards transformations of the agricultural sector. Land seizures in Mexico and passionate protests by the American Agriculture Movement suggest that dislocations have not been altogether avoided, nor has agriculture been successfully insulated from the economic, political, social, or climatological factors that have always affected it and the rest of the national community. In short, North American liberalization of trade is likely to be a derivative of domestic pressures and needs, rather than the result of policies directed specifically to that end.

Regionalism and Food Policy Futures

Given our regional approach to national food policies, what are the outline and prospects for future North American food policies? First, adjustments in either national or regional food policies will be inescapably predicated on interdependence. That interdependence, however, is largely informal and seems destined to remain so for the immediate future. Coordination, rather than integration, seems likely to prevail. As Keohane and Nye point out, informal coordination has become the customary pattern of relations between Canada and the United States.[44] The essence of that coordination is management through information exchange, rather than control by directive.

It is upon this foundation that regional agricultural policies must be constructed. Informal coordination is evident in the programmatic and structural similarities found in Canadian and U.S. agricultural policies. A somewhat similar coordination exists between the United States and the north Mexican agricultural sector, although the agri-economy and the agri-polity of southern Mexico are almost antithetical to that of northern Mexico, Canada, and the United States. Even in the underdeveloped south, however, the exchange of braceros for U.S. dollars represents a de facto, informal coordination with mutually beneficial results.

Therefore, in pointing to regionalism in national postures toward agricultural trade, we suggest that the benefits of their coordination could outweigh their competition vis-à-vis current potential trading partners. For example, Japan has become "Canada's second largest export market, after about 1973."[45] A regionally coordinated agricultural trading posture might conceivably influence Japan to reduce trade barriers, thus opening both agricultural and industrial markets for North American products. (See David Balaam's discussion of Japan's trade dilemma in chapter 6 of this book.) The bilateral, let alone trilateral, mechanisms for such coordination are presently nonexistent. Conceivably, internal institutions such as the American Commodity Credity Corporation, the Canadian Wheat Board, and Mexico's *Banco de Comercio Exterior* (Export Bank) could become the vehicles for so influencing trade.

Canada and the United States are widely regarded as major competitors in the international food markets. As such, it is argued, "the United States and Canada do not coordinate their food export policies in pursuit of common purpose."[46] Retrospective lack of coordination does not commit either nation immutably to such a deficiency. Indeed, it is the argument of this chapter that, especially with respect to the United States and Canada, programmatic and structural similarities are examples of incentives for increased management and coordination of agricultural policies in world markets.

The leap from discussing the possibilities of coordination of agricultural policies to the eventual likelihood of using food as a "weapon" should be avoided. Although the United States seems to have achieved some limited success in imposing and coordinating a grain embargo against the Soviet Union, these results are at least as significant for their symbolic worth as they are for their economic impact. Given the opportunity, Mexico would probably resist similar interventions in the internal affairs of a sovereign nation with whatever regional or world influence it could muster. Such resistance would diminish the symbolic impact of the "weapon." Using food as a counter to the oil cartel would, again, find the oil-rich Mexican government recalcitrant.

Nevertheless, even a loosely coordinated regional agricultural policy would provide enormous political influence for the coordinators. The impact of this influence probably would not be felt evenly across the entire spectrum of international affairs, but it could be important in areas of economics and, particularly, agro-economics. North American influence could even be used to bring some "rationality" to European and Japanese agricultural sectors. Food as a weapon would be regarded with repugnance by all members of the North American region, but food as a positive influence to world development might be regarded somewhat differently.

The eventuality of regional food coordination will evolve (if ever) from the reality of shared national interests. The use of food as a weapon by extension simply does not follow. In fact, regional coordination would probably inhibit the unilateral use of food in such a role because of the lack of uniform interest in the intended "target."

Seeking to use food as an influential element in interregional relations rests on common interests. For all three nations, agriculture is a source of foreign exchange and makes a positive contribution to national balances of trade. The foregoing examination of agricultural policy suggests that other similar policy goals exist. With respect to aid programs to LDCs, the nations of North America have similar interests in providing markets for agricultural exports as well as being concerned about promoting the process of nation building. Coordination of the pursuit of such goals could entail aid in the form of agricultural research and information suited to building an infrastructure for national food policies.[47] While it can be argued that such a policy is against the short-term interest of either or both Canada and the

United States, it should be pointed out that the underlying issue here is the maintenance of farm income and political support from agricultural interests. Certainly both short- and long-range interests in Mexico would be served by such arrangements.

Perhaps the strongest argument in favor of regional food policy coordination is made in light of trends in international trade. While protectionism has long been a dominant characteristic of that trade, regionalism and market management are comparatively recent. A case in point is the Common Agricultural Policy (CAP) of the European Community. North Americans agree in principle that the world's major importing region is unnecessarily restrictive toward their individual export overtures. Mexico has found the same problems with its northern neighbor. Nonetheless, they unilaterally make little progress in influencing the relaxation of quotas and tariff and nontariff barriers by a regionally unified community of nations. CAP's very success suggests the desirability of emulation.

However compelling economic arguments in favor of regionally coordinated policies may be, political arguments remain decisive. Arguments from autarky, nationalism, and emotion carry powerful weight in the councils of Canadian and Mexican government. Incremental and informal approaches, if successful and if accompanied by insurance of sovereignty, seem essential to the development of a closer continental community.

From a regional perspective, the problem rests in the yet to be identified forms and mechanisms through which an accommodation of varying trading styles and sensitivities can be effected. This means that each narrow issue area, commodity, and product, demands (at least for the interim) a uniquely developed and justified "mini-policy agreement." Such agreements must be absolutely predicated on the political and economic realities of the actor states.

Given existing degrees of interdependence, programmatic and structural similarities, complementarity, and a rapidly changing set of international global alignments, the question becomes less one of probability than of timing, tactics, and opportunity. Regional coordination and management are likely to occur when they are perceived to be politically desirable and economically beneficial. While the bilateral, incremental moves suggested here are certain to have multilateral consequences, it is difficult to foresee them. Certainly some will be unintended, some serendipitous, and some counterproductive; only ongoing accommodation and adjustment among the constituents of North America can address these eventualities. The successes of the Western and Eastern European economic communities provide both challenge and inducement for the nations of North America.

The superpower role of the United States almost certainly will have a debilitating effect on efforts toward more harmonic economic coordination in North America. The strategic interests of Canada and Mexico are not in all instances in complete accord with those of the United States. For example, Mexican and Canadian relations with China and Cuba often charted courses more in the direction of their own self-interests than American security concerns.

It is precisely in the strategic area that regionally coordinated management might find some of the best and worst prospects. Together Mexico, Canada, and the United States would enjoy a stronger hand in GATT and bilaterally with the EC/CAP. More trade would ensure agricultural markets and continuing structural adjustments in respective domestic sectors and promote global approaches to shared problems. On the speculative side, Canada and Mexico, with their vast energy resources, might be more inclined than the United States to nonintervention in the Middle East. It is doubtful, however, whether this would directly affect the issue of food policy, since it is widely agreed that food is not as effective as petroleum as a strategic resource.[48] Food and food policy, to repeat, are construed here as a resource of influence, rather than a "weapon" of compulsion. Hence, with regard to food policy, to repeat, Canada, Mexico, and the United States might practice regional coordination in the absence of consensus on strategic issues.

Conclusion

Regional coordination of food policy is no panacea for the ills of mankind. Its scope is limited. It alone cannot alter the military balance. Further, structural adjustments in national agricultural sectors cannot be prevented, although they might be mitigated by collective domestic pressures. Further rationalization in industrial trade issues, government role versus market role, will probably be pursued independently. The basis of food supply to LDCs is not likely to change. Cash and concessions will continue. Basic rationales for national food policies will persist, and regional food policies will be in large conformity with these. Regionalism will not ipso facto address fundamental moral questions associated with human nutrition.[49] The influence of food is most likely to be felt in developed countries' trade relations, not in developmental schemes associated with LDCs. Regional food policy can have little impact on the cultural factors that impede the achievement of self-sufficiency in some LDCs. Fears of domestic agricultural interests about loss of export markets appear to conflict with making those same markets self-sufficient.

Domestic factors will continue to dominate the evolution of regional coordination. In Canada and Mexico, nationalist and separatist attitudes make the prospects for regionalism appear difficult. The geopolitical aspect of these sentiments—commercial agriculture in the prairies and northern Mexico, subsistence agronomy in southern Mexico, fisheries in the maritime provinces, and industrial development elsewhere—promise parliamentary and intraparty conflict the results of which cannot be anticipated.

All the foregoing notwithstanding, the advantages of regional coordination of food policies seem to outweigh the disadvantages. Historical patterns of Canadian, Mexican, and American agricultural development are as regionally compatible as are other nations' experiences, especially Europe and the CAP, who likewise face global uncertainties. Programmatic and pragmatic similarities in the region are striking, as are the most extreme disparities. Agricultural policy is a component of interdependence. Trends in international food trade reflect increasing governmental involvement.

The rationalization and incremental improvement of interdependence in the area of food policy will occur, if at all, not "with a bang but a whimper." It will not bring the millennium. It could improve national, regional, and global food situations. It will not eliminate hunger. It certainly will not bring the political unification of North America. Perhaps, its greatest effect would be exemplary. This regional coordination, if successful, might inspire similar efforts in Asia, Africa, and South America. Given the near complete disorder of the world food situation, such efforts of rationalization could only be welcomed.

Notes

1. Dennis Pirages, *The New Context for International Relations: Global Ecopolitics* (North Scituate, Mass.: Duxbury Press, 1978).

2. Richard J. Trethaway, "International Economics and Politics: A Theoretical Framework" in *The Interaction of Economics and Foreign Policy*, ed. Robert A. Bauer (Charlottesville: University of Virginia Press, 1975). See also John Zysman, "The State as Trader," *International Affairs*, vol. 54, no. 2 (April 1978), and J. H. Richter-Altschaffer, "Agriculture as a Problem in the Relations between Europe and the United States," *The Washington Papers*, vol. 2, no. 2 (Beverly Hills and London: Sage Publications, 1974).

3. Roy M. Robbins, *Our Landed Heritage: The Public Domain, 1776–1936.* (Lincoln, Nebr.: University of Nebraska Press, 1963).

4. See, for example, James E. Anderson, David W. Brady, and Charles Bullock, *Public Policy and Politics in America* (North Scituate, Mass.: Duxbury Press, 1978) and Theodore J. Lowi, *The End of Liberalism: Ideology, Policy and the Crisis of Public Authority* (New York: W. W. Norton Co., 1969).

5. Vernon C. Fowke, *Canadian Agricultural Policy: The Historical Pattern* (Toronto: University of Toronto Press, 1946).

6. Organization for Economic Cooperation and Development, *Recent Development in Canadian Agricultural Policy* (Paris, 1978), p. 4.

7. Lester R. Brown, "The Next Crisis. Food," *Foreign Policy*, no. 13, (Winter 1973/74), p. 21.

8. *Latin American Weekly Report*, vol. 1 (2 November 1979), p. 6.

9. Anderson et al., *Public Policy*, pp. 382–83.

10. OECD, *Recent Development*, p. 43.

11. Ibid.

12. Anderson et al., *Public Policy*, p. 381.

13. T. K. Warley, "Canada, Australia, and New Zealand in International Agricultural Trade," *U. S. Agriculture in a World Context: Policies and Approaches for the Next Decade* (New York: Praeger Publishers, 1974), p. 112.

14. OECD, *Recent Development*, p. 45.

15. Ibid., p. 29.

16. OECD, *Recent Development*, p. 33.

17. Gary L. Benjamin, "Minor Changes in 1979 Feed-Green Plan," *Farmer's Digest*, vol. 42, no. 8 (February 1979).

18. Anderson et al., *Public Policy*.

19. Travis W. Manning, "The Agricultural Potentials of Canada's Resources and Technology," in *Readings in Canadian Geography*, ed. Robert M. Irving (Toronto: Holt, Rinehart and Winston of Canada, 1978), pp. 144–49.

20. D. Gale Johnson and John A. Schnittker, eds., *U. S. Agriculture in a World Context: Policies and Approaches for the Next Decade*, (New York: Praeger Publishers, 1974).

21. Manning, "Agricultural Potentials."

22. Benjamin, "Minor Changes in 1979 Feed-Green Plan."

23. Lowi, *The End of Liberalism*.

24. Charles B. Shuman, "Food Aid and the Free Market," in *Food Policy: The Responsibility of the United States in the Life and Death Choices*, ed. Peter G. Brown and Henry Shue (New York: Free Press, 1977).

25. David Balaam and Michael Carey, "The Influence of Domestic Constraints on Foreign Policies Regarding the Issue of Food: Comparison of the U.S. and EC," unpublished paper, Western Political Science Association Meeting, Los Angeles (March 1978). See also Francis E. Rourke, *Bureaucracy, Politics, and Public Policy*, 2d ed. (Boston: Little, Brown and Co., 1978), and Theodore J. Lowi, *The End of Liberalism*.

26. Glenn C. Lorang, "Would a 'Wheat Board' Get More for Your Crop?" *Farm Journal* 102, no. 6 (April 1978): 26.

27. *Latin American Political Report* 13, no. 35 (September 7 1979): 277.

28. Robert O. Keohane and Joseph S. Nye, *Power and Interdependence: World Politics in Transition* (Boston: Little Brown, and Co., 1977).

29. Ibid., p. 209–10.

30. Ibid.

31. *Latin American Political Report*, p. 277.

32. Keohane and Nye, *Power and Interdependence*, p. 214–16.

33. Ibid., p. 168.

34. Octavio Paz, *The New Yorker*, 17 September 1979, p. 150.

35. David L. MacFarlane and Lewis A. Fischer, "Prospects for Trade Liberalization in Agriculture," in *The Impact of Trade Liberalization*, ed. H. Edward English (Toronto: University of Toronto Press, 1968).

36. See James Trager, *The Great Train Robbery* (New York: Ballantine Books, 1975) and Gary L. Seevers, "Food Markets and Their Regulation" in Hopkins and Puchala, *The Global Political Economy of Food*, ed. Raymond F. Hopkins and Donald J. Puchala (Madison, Wis.: University of Wisconsin Press, 1979).

37. Macfarlane and Fischer, *Prospects for Trade Liberalization;* OECD, *Recent Development*.

38. Macfarlane and Fischer, *Prospects for Trade Liberalization;* Canadian-American Committee, *A Possible Plan for a Canadian-U.S. Free Trade Area* (Washington, D.C.: National Planning Association, 1965).

39. *Latin American Political Report*, p. 277.

40. Shuman, "Food Aid."

41. Seevers, "Food Markets."

42. See I. M. Destler, "United States Food Policy 1972–1976: Reconciling Domestic and International Objectives," in *The Global Political Economy of Food*, ed. Raymond F. Hopkins and Donald J. Puchala, *International Organization*, vol. 32, no. 3 (Summer 1978); Anderson et al., *Public Policy;* and Balaam and Carey, "Influence of Domestic Constraints."

43. Anderson et al., *Public Policy*.

44. Keohane and Nye, *Power and Interdependence*.

45. OECD, *Recent Development*, p. 25.

46. Robert Paarlberg, "Food, Oil and Coercive Resource Power," *International Security*, vol. 3, no. 2 (Fall 1978).

47. Raymond F. Hopkins and Donald J. Puchala, "Perspectives on the International Relations of Food," *International Organization*, vol. 32, no. 3 (Summer 1978).

48. Robert Paarlberg, "Food, Oil and Coercive Resource Power."

49 Brown and Shue, *Food Policy*.

3

European Food Policy: Rules of the Game

MICHAEL J. CAREY

This study deals with European food policy on a regional basis. It does so by examining multilateral European agricultural trade relations with major partners. It is at this level that the international rules of the game governing trade are created. At the same time, this study is concerned with partners on the bilateral level in the Mediterranean area. On one level, rules of the game become criteria or, hopefully, habits for the conduct of purely international economic relationships. The political content of the rules primarily concerns wealth and welfare issues for the rich regions or nations involved. On the other hand, at a second level, rules of the game not only include the wealth and welfare concerns of Europe and the Mediterranean, but impinge upon European security and ideological concerns. Europe's role in the world food arena is undergoing change; evolving multilateral and bilateral rules of the game demonstrate this point. As bipolarity declined and interdependence increased, new opportunities for an enlarged European role arose. One critical opportunity area is agricultural trade; but it is an aspect of trade more constrained than most by domestic demands. As domestic demands, international opportunities, and restraints change, European decision-making (especially in the European Community) can influence new patterns of trade, effect new temporary trade-offs between agri-trade partners, and create new international political solutions to trade (and domestic) conflicts. These patterns, trade-offs, and solutions constitute new rules of the game bearing a European imprint.

Regional food policy in Western Europe, specifically that of the European Community (EC), encompasses a host of issues. Each creates obstacles for decision makers; so much so that, in fact, the Common Agricultural Policy (CAP) is constrained by severe tensions. These issues involve economic, political, and strategic considerations that challenge the very idea of a world role for the EC. Hence, these modernized polities are faced with internal and external obstacles that make them incapable of governing and planning ahead in food policy. One important need is for new rules of the game governing trade relationships. Many policy areas today cross the boundaries of states, confront opposition and dissent, and affect foreign policies and domestic policies of other states. Agriculture and food policies constitute issue areas that are especially susceptible to the crises of interdependence, particularly in the EC. One set of domestic constraints involves interest groups affected by outside competitors; these groups generate demands for protection and are able to pressure governments as well as the EC for protection. These demands often reflect a general orientation toward "development" policies— i.e., agricultural, industrial, and often regional policies as well. Moreover, states often use agricultural policies to promote development and protect resources or aesthetic values in "special" regions, e.g., Brittany, rural districts in the United Kingdom, and Gaeltacht regions in Ireland. Values incorporated in various developmental policies advocated by groups, national governments, and the EC have become part and parcel of CAP, and have in turn become powerful constraints upon alternatives evolving within the CAP despite powerful outside demands (e.g., by U.S. decision makers and interest groups) for reform. On the other hand, this group involvement applies equally to fisheries policies as well as the CAP. Developmental prospects for new members, e.g., Ireland, are encouraged by disappearance of North Sea herring, discovery of Donegal herring stocks, and rapid expansion of traditional onshore fisheries into offshore exploitable resources. Consequently, interest groups grow more protectionist.

Interdependence

The key political question facing decision makers is: can an agriculturally interdependent world be "managed," and if so by whom? Managers will have to be national and regional actors, i.e., decision makers, and in rare cases nongovernmental decision makers, since there is no other source of "legitimate" management in international relations today. The alternative is, of course, that interdependence cannot be politically managed and the chaos of international wealth and welfare will continue and probably intensify. Strains between allies in the West and peripheral partners will reach a breaking point: more specifically, principals of GATT will be ignored, protectionism will increase, or regional blocs will be created. If interdependence cannot be managed, internal tensions will also increase as domestic groups raise their demands for protectionism. Since not all domestic groups will benefit in the short run from protectionism, political conflict within polities may increase as well. Even though larger groups (such as consumers) who will suffer from protectionism are unorganized, some political reaction to the shortsighted demands by highly organized interest groups can be expected. Overall, the result of nonmanagement of interdependence will be internal and external political instability.

The key problem of interdependence of political decision makers is management of or creation of rules of the game. Control is out of the question now; interdependence has developed past the point of control by large and medium powers. Even foreign policies specifically aimed at autonomy, e.g., those of Gaullist France, were bound to fail insofar as the process of change in the international system exceeded any one state's ability to control it.[1] The very concept of interdependence implies a lack of directability by national and regional actors since the locus of activity is not simply within a state or region, but occurs within the interactions of several or many states. Moreover, where decisions by officials are ineffective, no control (in the narrow sense) is possible. Decisions facing key decision makers are increasingly ineffective since the activity occurs simultaneously within states, between states, through regional actors, and perhaps through nongovernmental organizations such as multinational corporations and other forms of transnationalism.

The states most affected by interdependence are the developed countries; consequently these states stand to gain or lose the most from interdependence. It is Western Europe, the United States, and Japan that dominate multilateral forums where negotiation on interdependence takes place (e.g., GATT). These states are in the forefront of regulation attempts designed to tame interdependence. However, regulation is distinct from control, since control implies long-term power over interdependence; yet interdependence, as is becoming increasingly apparent, can at best only be managed (if that), and not dominated. Management of interdependence by these states means reaching some kind of agreement on principles of cooperation or rules of the game, by which political and economic dislocations can be minimized internally, some rules for nongovernmental organizations established, behavior appropriate for crises determined, and benefits of interdependence spread to other regions. Trade relations are the logical starting point for rule making. International trade regimes created in the past have thus increasingly become forums for quiet negotiation of new rules and arenas for intense struggle. Agricultural and industrial trade both constitute areas for this deliberation and decision making.

"Countervailing power" (through creation of blocs) for protection or continued emphasis on self-sufficiency may well mark the new strategies for coping with interdependence.[2] Rules of the game and cooperative solutions, however, are more likely patterns of problem resolution than are solutions like autarky or the development of countervailing power. Indeed, interdependence has developed to a point where previously separate problems have disappeared due to their merging with broad, interconnected problems in both agricultural and industrial trade. Discrete solutions increasingly fail to work. Ideally, interdependence requires joint efforts, pooled sovereignty, etc. Yet, "nations are loath to surrender their autonomy. Sharp differences in the analysis of the causes of problems and in preferred solutions are the essence of international economic disharmony."[3] Compartmentalization of institutions organized to solve problems often further limits responses.

At the same time, little operating room exists for the notion of control in interdependent relations. A narrow conception of economic power implies that a state's or region's economic policy can be used to modify others' behavior.[4] Such a conception employs a simplistic "carrot and stick" metaphor, and ignores the situations that require interaction by key decision makers to establish mutually advantageous environments, or to limit negative effects from which many may suffer. That is to say, effective decision makers must increasingly consider not only other nation-states' policies, but also the effects (both external and internal) of their decisions. While the ideas that "everything is in flux" or that "all is connected" do not help much in clarifying interdependence and the decision makers' dilemmas, those ideas do impart the flavor of interdependence and the plight of the decision maker who finds himself in the middle of it. More groups participate whether or not decision-makers desire it (e.g., COPA, other green forces, and AFL-CIO), domestic policy debate widens, and new complexities arise that defy rational categorization, e.g., which department or ministry should deal with them. Examples of widening domestic policy debates are found in European food policy in regard to surpluses of beef and milk products, and development of export markets desired by producers versus budgetary constraints that some members would like to see imposed on CAP. Rational categorization examples are found in Agriculture Ministries versus Foreign Affairs Ministries versus the Exchequer debates on CAP within member states, and with the Commission.

Fortunately, strong incentives exist for decision makers to rely not solely on attempts at control, and narrowly defined power conceptions, since long-run interests will be served by finding alternatives to trade wars. Rules of the game (via EPC) are made possible by (1) the nuclear balance of terror and (2) economic diplomacy. The first point eliminates whole areas and styles of action for Europe. The second reveals an enlarged arena for Europe, especially since the collapse of the Bretton Woods system signaled the end of absolute U.S. hegemony in the developed world. Hence, new rules of conduct were, and are, being written. Today that means EPC can follow several courses. Two of the most important concern U.S.- European relations and European-Mediterranean relations. The first set tends to assume conduct-determining importance for the developed world. The second involves examples of resource supply, market, and security restraints. These degenerated during the oil crisis of 1973–74 into beggar-thy-neighbor and decidedly un-European actions by members of the EC. Nevertheless, bilateral relations have been of the utmost significance.

The EC appeared to resolve some of the dilemmas posed by interdependence by accepting the principle of an internal food policy regime codified in the CAP during the 1960s and early 1970s. The member states (the Six) chose to "institutionalize interdependence" and create an integrated food market. The French-German trade-offs and other details of the birth and evolution of CAP are recorded elsewhere.[5] The members had foregone two alternative methods of dealing with interdependence: either going it alone, or relying upon a broadly based "and coordinated action among the industrialized societies of the non-Communist world."[6] Both narrow nationalism and super-GATT alternatives were rejected; rather, the integrative, regional approach was taken. While integration proceeded haltingly, interdependence proceeded quickly and generally disrupted domestic, foreign, and regional politics in Europe and elsewhere. In the EC, backsliding in the face of these pressures occured, and various member states occasionally returned to national measures, especially in food policy problems. Generally, though, even with its problems the CAP stands as the best example of regional integration in Europe; so much so that members view its existence as nonnegotiable in GATT or other fora.

For the Europeans, regional food policy is hardly so simple a matter as the forging of a CAP and its structures; rather, regional food policy reflects a full range of cross-cutting political interests of the Community, including security, ideology, hegemony, trade, and resources. This chapter poses the dilemmas of regional food policies from a broader perspective of interdependencies, examining interdependence not in the narrow terms of trade flows, but in terms of the often contradictory and always formidable political dilemmas of Europe's special place in an interdependent world. Relying mainly on civilian power and European Political Cooperation, European decision makers within the EC struggle with domestic demands and foreign policy requirements. Europe is limited by its middle position relative to the United States and the Soviet Union. Since it lacks effective military strength it must rely instead on its political and economic resources. Europeans gropingly find their way in a world still dominated, even in détente, by two superpowers. Europe may challenge either one politically and economically but not strategically. By using civilian power, Europe creates a new role a portion of which is defined by domestic requirements; e.g., internal food policy vis-à-vis the United States, where conflicts of interests look like simple zero-sum situations. Further, Europe's role is being defined in terms of the Middle Powers' relation to nearby areas (e.g., Mediterranean areas, the Third World), by enlargement of the EC, by problems such as creating a liberal order of world trade, guaranteeing access to resources and markets, and even insuring equal development within the EC via monetary union (EMU). The creation of EMU could create a "rival" to the IMF, and may be the beginning of a currency bloc dividing the West. EMU would alter the face of CAP by reunifying the market within the EC.

The European Community

The EC operates in cycles of despair and optimism: despair during those movements or years of downturn in the integrative process and optimism with each relaunching of Europe. The most recent cycle is the ten-year period of 1969–79. The upturn of 1969 marked by invitations to the UnitedKingdom, Ireland, Denmark, and Norway, talk of an Economic Union, and steps to create a common foreign policy for the EC (now referred to as European Political Cooperation) turned into despair as enlargement of the EC produced neither economic union nor a common foreign policy after 1972. Oil supply troubles, exchange rate difficulties, changes in America's foreign policy, and neomercantilistic trends in the world economy all reduced integration in Europe to a shambles.[7] The down cycle of despair was reversed by emerging integrative policies relaunching Europe: the emergence of EMU, now known as ECU (European Currency Union), scheduling of direct elections to the European Parliament, institutionalization of EPC, and a sense of confidence in new rules of the game—largely the result of EPC successes by 1978–79. The euphoria of European decision makers, however, is less over ECU than over the "relaunching of the EC."[8] Clearly, for the new monetary system to work, closer economic cooperation is necessary, and its success will simplify a return to a common internal food regime. Perhaps less spectacular but more important, the creation of the EPC clearly signals an upturn in the cycle since EPC may have greater long-term effects. This is also an area of utmost importance to all other actors in the international system, large and small alike.

EPC concerns outside actors since it means that the nine nations coordinate their foreign policies, and at times act in complete unity. While Europe of the Nine often speaks with one voice, this is nowhere more apparent than in regard to food policy. EPC created new rules of the game for European decision makers by allowing less controversial issues (and even some defused controversial issues) to be pursued in international forums at the multilateral level (such as GATT).[9] Further, the Europeans may be even more united in "bilateral" relations, that is, EC–third country relations that fall outside the realm of GATT, which are often relations in which security concerns provide the substructure to superficially economic relations, e.g., relations with Iberian nations. Other bilateral relationships may present more of a patchwork of intentions, but nevertheless indicate new roles for Europe and new rules of the game. "New rules of the game" (as the term is employed here) concerns both new patterns of interests and solutions to problems of interdependence, although it does not necessarily imply either harmonious or cooperative behavior among all actors. The latter may be necessary to a "best for all" solution. However, developed nations with many common interests cannot work out rules of the game continually advantageous to them. Accordingly, new rules of the game are reflected in two basic patterns negotiated by the EC via EPC at both the multilateral levels (GATT), and bilateral (Iberia and the Mediterranean) levels.

Before analyzing EC regional food policies specifically, European agriculture must be placed in its world context. The EC maintains a higher proportion of the work force in agricultural occupations than is common in other developed nations or regions, largely due to a number of conscious political decisions the net effect of which is to slow the rate of departure from agricultural employment to industrial and urban employment for social, political, and even aesthetic reasons. Compared to other developed regions or nations, the EC is largely self-sufficient in most temperate-climate agricultural commodities. The EC is, however, the largest food importer and second only to the United States as food exporter. The EC imports more food than it exports, to the tune of a farm trade deficit of more than $21.5 billion in 1975, quite a contrast to the U.S. agri-trade surplus of $13 billion in 1975. Price advantages, feed grain shortages, and need for tropical products account for the EC's import position. For example, in 1975 the EC purchased $5.6 billion worth of U.S. farm produce, which was five times more than the EC sold to the United States. The 1975 farm trade surplus for the United States was more than double the total agri-exports to the EC in 1972. In 1977 the U.S. surplus in agri-trade with the EC was $4.8 billion. For comparison's sake, previously the United States sold $6.2 billion worth of farm exports to all countries in 1968.[10] The EC, as the largest importer in the world, takes up 35% of all farm products traded on international markets.

In regards to developing nations, the EC both imports and exports, as well as provides food aid and agricultural expertise. As with most rich countries, the emphasis is on sales. The EC share of world agri-exports rose to 12% of the total in 1975. This, it is claimed, is "helping to increase food security in the world."[11] Critics of European food policies, on the other hand, argue that imports from the poor nations will not grow fast enough to promote the general welfare of the poor countries. In regard to the locus of aid, the EC gave nearly all of its support to Mediterranean, African, and a few Asian states. More specifically, most EC aid (all types) went to associated states and territories that maintain some traditional link to European states as a holdover from colonialism.[12] Roughly $100 million was spent on the food aid cereals program in 1976.[13] More importantly, expertise is also exported unilaterally to Third World countries[14] by various EC member states and through EC programs. If such aid contributes to the development of indigenous agricultural capabilities, its impact will be far greater than simple food aid.

In general, reflecting the complex patterns of interdependence previously summarized, EC agricultural policy affects many other food producers in the world. The premises behind EC policy differ little from the premises of other countries' policies. However, policies do differ, with regard to the clarity with which EC policy is declared. Unlike the United States, with so many voices clamoring to speak for agriculture, in the EC generally one voice is heard internationally, and that is the voice of the EC Commission (under the guidance of the Council of Ministers). For example, the Commission spells out the premises as follows: feeding the population of a nation or group of nations is indispensable to independence. Food supplies are then secured through a region's own agricultural resources or external trade. Neither source is absolutely certain nor always available due to productivity endowments and the "unorganized state of world commodity markets [that] prevents any logical distribution of surpluses or shortages occurring . . . under the influence of weather, speculation or even politics."[15] Hence, the EC pursues particular agricultural policies that "cannot be set on the same footing as manufacturing industry where the problems faced are of a very different kind."[16] Due to various natural and human factors, then, agriculture tends to be supported by government intervention "in ways which are, to varying degrees, direct and open."[17] The EC accepted from the beginning the principle that agriculture could not be regarded as fulfilling a role comparable to that of industry in creating a common market. In practice the agricultural sector became the most successful and expensive policy arena resulting from the Treaty of Rome. Internally this meant a policy consistent with the goals of: (1) securing supply, (2) insuring "reasonable prices," and (3) guaranteeing a common internal market without barriers. Various pieces of EC machinery evolved to manage the agricultural sector, the most important of which is FEOGA (the guidance and guarantee fund). More importantly for the world, the CAP's external policy was what might be termed one of "protection without closure." As members of GATT, FAO, OECD, and other inter-national organizations, the members of the EC must foster liberal trade policies. However, agriculture (usually the unhappy stepchild of inter-national trade) retains its special protected status in the external policy of CAP (and the other developed nations). The problems these contradictions in food policy create are examined below under the discussion of GATT. The preceding trade statistics point to the fact that while the EC jealously protected its farmers, it also increased imports—especially from the United States. Despite the facts that some commodities cost more in the EC than in the United States, that subsidies continue, and that surpluses are occasion-ally dumped, the CAP appears to have generally promoted food trade growth. No developed country is willing to completely free its trade in food, for the very reason the Commission cited.

Despite the loud complaints and difficulties encountered in GATT with the United States and other developed countries such as New Zealand, the most serious problems with agri-trade may very well concern EC bilateral relations with the Mediterranean area, especially if the EC is enlarged to include Greece and Iberia. These relations, again, extend far beyond mere trade, and impinge on the very conception of CAP, perhaps even to security concerns of Europe.

From a free trade perspective, difficulties within the CAP would include the following: CAP was created with little attention to "integrating Com-munity agriculture in the world economy";[18] it is, accordingly, unable to cope with monetary disequilibriums, and tends to encourage food sur-pluses.[19] All of the critiques of the CAP, however, suffer from the flaws endemic to proposals for the rationalization of agriculture. Agriculture defies rationalization, due to the factors pointed out by the Commission. More than most other economic processes, the agricultural sector is affected by both environmental influences (such as the vagaries of the weather), and human influences (such as the desire to guarantee supplies). The only guarantee of enough is to supply too much! Predicting gluts and shortages is so difficult that most analyses of CAP become dated within a year or two (often less).[20] Protein shortages due to anchovy harvest failures in the Pacific off Peru, weather- and mismanagement-induced wheat shortages in the U.S.S.R., or milk gluts in Europe or the United States persistently play havoc with the best-laid plans of men. The difficulties of rationalization are aptly described by an EC official:

The dividing line between having a sufficient supply of any food stuff, a reasonable price, and having a glut with prices collapsing is quite small. Even 1% oversupply will cause a market to collapse overnight, and in the absence of a mechanism for withdrawing that extra-supply producers can suffer heavy losses which will eventually force them out of business leaving the market undersupplied and prices soaring.[21]

As the current Commissioner for Agriculture in Brussels, Finn Gundelach, suggested, the CAP provides food supplies for 250 million people "in an uncertain world at fairly stable prices, sometimes lower than on the world market."[22] Indeed, even the commissioner in charge of the budget, M. Fugendhat, who thinks CAP is too expensive, values price stability and security of supply as "important enough to justify paying a premium in terms of the actual level of prices."[23] In terms of "reform," the 1978–79 program increased farm prices only by an average of 2.25%, while simultaneously recognizing the significance of the EC to the Mediterranean area by attempting to harmonize agricultural policy within the EC to promote development in peripheral regions that fear competition from Greece, Spain, and Portugal. In addition, the reductions in complicated monetary compensation amounts (MCAs) may end the disruption of the common market caused by adjusting for fluctuating exchange rates.[24] Such rationalization of agriculture may not satisfy all economic experts but represents feasible change given the restraints of agricultural policy.

The Multilateral Level

The main protagonists of Western-world politics are the United States and Western Europe. This is nowhere more clear than at the multilateral level of EC relations, where stand-offs between the United States and the EC have been the major theme since the Kennedy Round negotiations in GATT. Today Japan receives more than its share of antipathy from domestic groups in both the United States and Europe, primarily labor unions and various declining industries. Nevertheless, it is EC-American relations that truly dominate the GATT forum, inasmuch as both are formidable actors in the international arena. Through EPC, European nations often pursue usually compatible goals, e.g., in agri-trade. Multilateral negotiations provide testing grounds for both parties—especially since 1971, when the U.S. hegemonic position was effectively challenged by Europe. The contest today is waged on a much more equal footing. Hence, it is unfamiliar to the United States, which was understandably happy with the earlier bipolar state of affairs in which Europe danced to a U.S. tune. Indeed, it is clear that U.S. support for the EC eventually provided the undoing of U.S. policy—as the EC became a successful competitor to the United States, and grew to a point where it pushed the United States out of its position of economic (though not strategic) hegemony. The decline of the United States (signaled by increased trade problems, currency difficulties, protectionist demands by interest groups, and strategic decline relative to the Soviet Union) prompted EC political demands for changed rules of the game reflecting EC growth. The United States created or dominated the rules of the game until about 1971, when an interrupted dialogue reflecting complex interdependence produced new rules of the game. While new rules have not been fully agreed upon by GATT members, this organization's role in facilitating trade and growth through the 1950s and 60s makes it look like the best bet available to solve trade problems of the 80s when doubts about the feasibility of continued economic expansion and fears of competition in the industrialized world, lead to protectionism and neomercantilism. Agri-trade traditionally receives the most concessions for protection and other forms of illiberal trading practices that have constituted the rules of the game since the Kennedy Round of GATT negotiations.[25]

The winter of 1979 finds GATT facing an unfavorable atmosphere: the EC faces countervailing duties on its agricultural exports to the United States tried to halt GATT negotiations in the face of this Congressional action, but the Commission and two large member states persuaded others to of the executive and oppositional attitudes in Congress.[26] Several EC member continue the negotiations. The American executive continues to seek in-creased markets for U.S. agricultural goods via tariff reductions, as well as better terms of trade for these products. The most important issue may well be not tariff reductions themselves, but rather the selective protection clause that will apply to individual countries. The American position links agricul-tural and industrial trade, attempts to get concessions on agri-trade from the EC, and attempts to justify the tariff concessions on industrial imports (steel, autos) that many American interest groups oppose. American arguments here are buttressed by America's comparative advantage in agriculture.

Some EC members interpret U.S. actions as an attempt to destroy CAP. The Commission and members stress the favorable agri-trade balance that the United States enjoys. While members differ on interpreting U.S. actions, agreement exists that they will take vigorous action to insure that CAP will not be dismantled. When countervailing duties are imposed by the United States, a dramatic and dangerous sequence of events is likely. The EC cannot avoid countermeasures.[27] Europeans consider U.S. countervailing duties unfair because they can be levied without any demonstration of adverse effects on American industry or agriculture. For America to impose countervailing duties, the U.S. Treasury has only to find that an import has been subsidized—and as noted above CAP relies on subsidies extensively as a policy tool.

The question of subsidies is especially difficult, and further complicates the process of generating acceptable rules of the game. Subsidies assume different significance in the eyes of various beholders. Subsidies reflect differing political attitudes domestically, which determine acceptable levels of governmental intervention in the management of the economy. American political culture, permeated by a myth of *antidirigisme*, influences executive-office negotiators in GATT. EC decision makers, while not guilty of ignoring government's role, perhaps elevated it to too high position in their polities. Nevertheless, just what constitutes a subsidy, and how it fits into liberal world trade, still vexes GATT negotiations. GATT does not prohibit the use of subsidies, but requires their control by notification and consultation. The 1970 GATT declaration contains an obligation "subject to certain conditions, not to apply export subsidies."[28] Further, if a subsidy threatens to injure a producer in the importing country, countervailing duties can be applied. U.S. legislation is not in line with this GATT rule, since any subsidized import is, even if it causes no injury, open to countervailing duties. Moreover, the United States wants to include regional domestic subsidies and other internal aids as liable to countervailing duties. Agreement on criteria for determining injuries and common interpretations of what constitutes a subsidy appear distant. Subsidies are a key to the operation of the CAP, so GATT partners will have to accept a definition of subsidies that allows CAP to work, and that will be quite uncomfortable for some U.S. interest groups.

If American-Japanese tentative agreements on agri-trade in GATT are any indication, the rules of the game are not in for much change, as the United States sought agri-trade access to Japan similar to that it now seeks in the EC. While some Japanese quotas on particular agricultural products doubled, very little was accomplished since the original quotas were so strict. Despite evidence to the contrary, U.S. negotiating officials made the concessions appear more important than they are, and, of course, agreement between all three major GATT partners (U.S.-Japan-EC) hardly moved closer. EC commissioners urged both the United States and Japan to make greater tariff cuts.[29]

For several reasons, the EC has sold 7 to 8 million tons of wheat in new world markets that it had not previously penetrated. Subsidies in the EC were raised for these exports, and the American crop was smaller in 1978; hence, prices in those world markets rose, and U.S. farmer-held reserves contributed also to making wheat scarcer. U.S. interest groups (like the wheat-exporting companies) are seeking a wheat export equalization program to maintain "their overseas markets."[30] The last U.S. export subsidy of this sort ended in 1972. The EC proposes a world wheat cartel (actually dominated by developed country producers) to solve specific price and supply problems; yet, generally, the United States opposes such *dirigiste* (interventionist) policies, while interest groups ask for them and/or for government support.[31]

More broadly, major multilateral EC goals are to (1) improve the provision of information between producing and consuming countries, and (2) reduce the excessive fluctuations of three important markets: cereals, dairy products, and meat products (especially beef). The point is to improve conditions in the market by increasing information on foreseeable supply and demand, and to solve access problems, such as the soybean embargo, by increasing dependability of supply—a factor often overlooked in the United States.[32]

Despite U.S. demands for access, American actions in Europe undermined their case for access to EC markets for agricultural goods. An embargo placed on soybeans and high-protein substitutes in 1973 brings U.S. reliability as a supplier into question.[33]

The Bilateral Level

EC bilateral relations (that is, with specific other countries or regions) demonstrate a whole range of food policies that are actually much wider in scope and implication than EC multilateral food policies. While the latter tend to be limited in scope to economic welfare, the bilateral relations go beyond economic welfare, and include strategic considerations and specific ideological values. Bilateral agricultural policies then become both ends and means for the EC at this level. These aspects of EPC and various other foreign policy goals (such as security) typify the cross-cutting issues that make agricultural policy an issue of larger concern than just to farmers, markets, and resources. Most clearly these objectives of EC food policy are observed in the Mediterranean area, and include (1) enlargement of the EC and (2) EPC policy toward other Mediterranean states. The issue of enlargement demonstrates most clearly the wide scope mentioned above. On the other hand, the so-called EPC global Mediterranean policy is narrower but not without political and strategic considerations as well.

The likely candidates for enlargement are Greece (scheduled to join in 1981 with a five-to-seven-year transition period), Spain, and Portugal. The leaders of both of the latter nations have pursued formal and informal discussion with the EC,[34] as did Greek officials leading up to the application decision. The main agricultural implications of this issue concern the nature of these regions' products and prices. The products of Greece, Spain, and Portugal not only compete with Mediterranean crops of France and Italy but, more importantly, undercut EC production by lowering prices. The key product sectors are wine, olives, fruit, and vegetables. Further, very large agricultural sectors exist in all three nations' economies.[35] In particular, Portugal and Greece resemble developing nations in terms of population employed in agriculture. Hence, to these prospective members, agriculture is critical.[36]

Interestingly, the CAP deals with most of the semitropical products in a fashion quite different from that adopted for temperate products in Northern Europe—in many ways the CAP protects, promotes, and concentrates on the northern farmer, and offers much less to farmers of the Mezzogiorno or Languedoc. These "domestic" effects of CAP are largely outside the scope of this chapter. The main point in the present context is that new competition for already peripheral farmers will result from enlargement. Indeed, enlargement will affect Mediterranean policy as well. Countries on the littoral, but not able to join the EC, produce many similar and competitive products that will face the common external tariff wall of the EC. Special arrangements (as described below) notwithstanding, these producers will suffer from enlargement.

The advantages and need for enlargement are more significantly political and strategic than economic, although the various arguments favoring customs union, economies of scale, and all the other benefits of integration are present as well, perhaps more so in industrial than in agricultural sectors. Political and strategic needs and advantages result from ideological values and defense needs. Ideological values of the EC "require" that democratic European states be allowed to join. While the stability of democracy in Greece and Iberia may not be completely assured, the EC welcome these nations' transformations from authoritarian systems towards liberal democracies.[37] Through EPC the members direct financial aid to Portugal to encourage the transition to democracy. These states do (or will) meet the political requirements for admission. More importantly, the EC sees enlargement as a way to guarantee and promote the democratization process already begun in these states. On the strategic or defensive level, the EC would prefer to keep all of these states strongly allied with Western Europe, preferably inside NATO, and certainly not allow them to become "neutral" or associated with the Warsaw Pact. Although NATO and EC are obviously separate (Ireland is not in NATO), it is hard to conceptually separate what is joined in the minds of Western Europeans, especially in the largest EC members. EPC includes no defense policies. NATO matters are compartmentalized and neither officially part of Council of Minister concerns nor part of the Commission's prerogatives. Yet, quite correctly, attractive economic and political inducements such as EC membership can cement the prospective member states "to the West" and promote the growth of a democratic political culture. Since ideological and strategic elements are so powerful, it is clear that contentious economic issues may recede. None the less, agriculture continues to play a role and may become an inducement to enlargment, and at the same time threaten the precarious balance among "green forces"[38] in the EC and other European interest groups.

It remains doubtful whether after enlargement a "southern" green front will emerge in the EC comprised of Mediterranean agricultural producers, due largely to disparate green interest groups in France and Italy.[39] Reaction by protectionist groups in both countries will guarantee policy changes in CAP *before* enlargement making cooperation with newcomers during and after negotiations unlikely.

From the newcomers' perspective, political and economic reasons for joining the EC probably balance in the short run; each state in general shares these reasons. International identity, an end to isolation, and preservation of democratic regimes appear to be the dominant political rationales for joining. On the economic level, it is geographical proximity, the basic economic economic systems now functioning, and previous association with EC that provide powerful reasons to join.[40] While at least one analyst suggests that specifically Greek agri-trade (and more generally Iberian agri-trade) poses less of a threat than often maintained, contrary evidence suggests that indeed agri-trade constitutes a key factor for old and new members. For example, no progress was made in harmonizing agri-trade between 1961 (when Greece gained associate status with the EC) and 1977. Of course, during the era of the junta, relations were stalemated by the EC. Also, Mediterranean producers in the EC already challenged the lack of market organization and structural policies in CAP aimed at their regions through the French and Italian governments and in Brussels. This was done with good reason: lower-priced identical or competitive products are produced in the three newcomer states. All the newcomers import temperate agricultural products, the sector best covered by CAP, and hence will help decrease surpluses produced by Northern European farmers. It has been suggested that the poorest newcomer, Portugal, is likely to be a net loser in terms of participation in CAP compared to what Portugal pays into the community budget.[42] Perhaps producers of temperate products will be the main beneficiaries economically; that would be in keeping with the pattern of benefits presently in the EC. One other associated food problem area for Iberia is fisheries; both nations have large fleets supporting a food industry that contributes a large amount of their own GDP. Unfortunately, this food resource increasingly divides the EC. New 200-mile limits, closure of Icelandic waters and other fisheries, disappearance of entire stocks (e.g., North Sea herring), lack of a community-wide allocation system acceptable to all members, and lack of policing and agreement at the multilateral level over fisheries management coalesce to create one of the most fundamentally divisive food resource issues in the EC.[43] Relations with Spain on the fishing issue have not been helped by anomic activity and political violence and disobeying withdrawal orders from French, British, and Irish waters, traditional fishing grounds for 500 vessels from Northern Spain.[44]

From the EC perspective, cognizance of Mediterranean agri-trade issues led to identification of problem areas, and an outline for new policy within CAP, as well as the Regional Fund and Social Policy. From the Commission's Mediterranean study two major conclusions were that agriculture in this region is more depressed, and of lower growth than the rest of the EC. A number of regions "where incomes are exceptionally low are facing problems not solely agricultural but bound up with the level of general economic development."[44] Market organization is correctly blamed for some of these weaknesses. Perishable crops have in some cases only the common external tariff to protect them at frontiers. Further, these crops lack well-structured market organization and price policy, which exacerbates difficulties. The Commission recommended that it is necessary "to launch a large scale drive for the Community's Mediterranean regions, bringing the range of Community financial instruments convergently to bear." Laconically, the Commission adds, "this is even more important in the context of the Community's prospective enlargement."[45] Enlargement got no other mention in this report, which is surprising inasmuch as the report was the result of fears of enlargement.

The 1978/79 CAP negotiations to determine internal prices included an important set of Mediterranean proposals that dealt with those peripheral regions of the EC, and not with price policy per se. Negotiations only included these proposals due to Italian demands that "increase in support prices for northern products were to be traded against a better deal for the south."[46] Enlargement prospects would have been limited by both France and Italy had these proposals not been included. The proposals contained measures to alter CAP in regard to most of the products commonly grown in the region. It is clear from these outcomes of CAP negotiations that enlargement-related competition problems are quite real. This led to an expansion of CAP in support regimes and structural policies to satisfy long-standing demands from peripheral farmers that had been long ignored. Such is the price the EC is willing to pay for enlargement. The EC touted these changes as a change in overall direction: "Community agricultural experts are eager to stimulate increased efficiency among farmers in the north and south, so that attention can be paid to harmonizing Community-wide agricultural production when Greece, Portugal, and Spain accede to the Community."[47] In other words, an internal adjustment process was begun within the EC to facilitate future enlargement.

The other half of Mediterranean policy concerns other states on or near the littoral with which, for strategic, resource, and market purposes, the EC is developing a special bilateral relationship. Agricultural policy again is a critical component of EPC in the region that is largely determined by "noneconomic" purposes. The EC is asserting a role for itself that it previously did not have, particularly in terms of new rules of the game. Primary strategic concerns here reflect the presence of the United States, the Soviet Union, NATO, and the interests of various local "powers" in the Mediterranean and Middle Eastern region.[48] Europe's strategic interests should be satisfied by NATO, and occasionally, unilateral policies, e.g., arms sales. The changing balance of power in the region is not of central importance to this paper, other than to note that this change encouraged the increasing role for the EC. Too, it must be noted that the EC role is one almost entirely of a civilian nature. Foreign ministers in the EC are only too willing to let military-strategic matters rest with NATO and the United States.

In terms of energy resources the key is obviously to guarantee oil supplies. The EC's conservation record is far superior to that of the United States, and the EC collectively should be able to face an embargo next time without as much disunity as in 1973. Other resources needed are various Mediterranean food products; however, enlargement of the EC will reduce import requirements from nonmembers. In terms of markets in the region the EC supplies industrial, service, and agricultural goods. The so-called global Mediterranean policy (as EPC toward this region is known) consists mainly of trade agreements for industrial and agricultural goods. Resource and strategic purposes remain submerged in EPC. These purposes are more met by technical assistance, financial assistance, investment, promotion of regional cooperation, and access for migrant workers. For the EC, oil is the most important commodity traded; for the region itself, agriculture exports are the most important items (except for the oil-rich). Some analysts doubt that these agricultural benefits granted by the EC, e.g., access to EC markets, will provide sufficient economic benefits "to make very close ties between the countries of the Mediterranean and itself (EC) tolerable in the long run."[49] Diversification, promotion of basic foodstuffs, and structural change is not being promoted by agricultural exports of the region to the EC.[50] Through CAP, green forces in the EC will probably resist such changes if it means outside producers are shifting from luxury or semitropical items to competitive products. On the other hand, the EC can contribute to stability in the region, but not in a strategic manner. "Given its attributes, it can render the greatest service by promoting social and economic development in the area."[51] Strategic and resource demands will most likely push the EC towards alteration and expansion of agri-trade in the region even after enlargement. The EC created a role in the Mediterranean, now that role will make demands upon EPC.

In conclusion, this examination of European food policies indicates that these policies indeed differ at multilateral and bilateral levels. Cross-cutting issue areas are most pronounced at the bilateral level, especially in regard to enlargement, while economic welfare is the key to multilateral agricultural policy. On the one hand at the multilateral level in GATT, it appears that new rules of the game are up for grabs, and probably will be strongly influenced by the EC since the other major partner (the United States), is really a partner now rather than a hegemonic ally. The CAP itself will be preserved, and the EC will continue to push for dirigiste solutions to problems of supply and demand in world agri-trade. Their position as a key importer gives the EC a curious kind of power, since they can justify their own illiberal agricultural policies by pointing to the size of their import-produced agri-trade deficit.

At the bilateral level in the Mediterranean, clearly the EC carved out a role for itself, part of which means enlarging the Community to include several new members. That is being done partly through modification of the CAP. New rules of the game in the region have not yet fully evolved; however, it appears that agri-trade concessions must be improved before the EC can play a substantial role.

Notes

1. This theme is developed in Edward S. Morse, *Foreign Policy and Interdependence in Gaullist France* (Princeton: Princeton University Press, 1973).
2. See Luther Tweeter and James Plaxico, "U.S. Policies for Food and Agriculture in an Unstable World," *American Journal of Agricultural Economics* 56 (2 May 1974): p. 367. See also Kathleen Patterson, "Keeping Them Happy Down on the Farm," *Foreign Policy* 36 (Fall 1979): 63–70.
3. Thorald Warley, "Agriculture in International Economic Relations," *AJAE* 58 (5 December 1976): 821.
4. Klaus Knorr, *Power and Wealth* (New York: Basic Books, 1973), p. 79.
5. See for example Miriam Camps, *European Unification in the Sixties* (New York: McGraw-Hill, 1966), ch. 1.

It should be noted that an alternative to inderdependence appears to exist in autarky, but is it feasible? Autarky is neither desirable (due to its political and economic costs) nor possible in consumer-oriented Western societies, where demand for imported goods will continue. With autarky, conflict greater than that which would result from mere protectionism would occur as consumers demanding better or cheaper foreign goods would confront protectionist domestic groups. Autarky is no answer if the problems of interdependence are global. For example: salmon and albacore migrate throughout the Pacific Ocean. Commercial fishermen from Oregon who compete with other nations' fishermen for salmon spawned in Oregon (or in other parts of western North America) cannot keep the fish under lock and key. Herring and cod in the North Atlantic present a similar problem. Other resources have similar problems: some resources cannot be found "at home," e.g., coffee. More generally, and perhaps more importantly, the overall benefits of "free" trade and communication appear to be so substantial as to make giving up interdependence too difficult and too costly. Decision makers and voters may well prefer economic dislocations plus benefits over autarky plus less choice. Consequently conflict will increase in domestic politics over who is hurt by and who benefits from interdependence.

6. Edward Morse, "The New Europe, a Unified Bloc or Blocked Unity?" in *The Interaction of Economics and Foreign Policy*, ed. Robert Bauer (Charlottesville: University of Virginia Press, 1975), p. 126.

7. See Morse, ibid., pp. 103–9 for an analysis of this period.

8. See, for example, *The German Tribune*, 1 November 1978, 866/26, p. 2.

9. The ambiguous and dynamic nature of EPC is analyzed in Michael J. Carey, *The Europeanization of Irish Foreign Policy* (unpublished Ph.D. dissertation, University of California, Santa Barbara, 1977), ch. 7.

10. These figures are from former agricultural commissioner of the EC Pierre Lardinois, *Newsletter on the Common Agricultural Policy*, September 1976, no. 7 (Brussels), p. 1; and see *The Week in Germany* 9, no. 69, (1 December 1978): 4 (Bonn).

11. Ibid. For a more comprehensive set of statistics see Simon Harris, "EEC Trade Relations with the USA in Agricultural Products" (Center for European Agriculture Studies: Wye College, Ashford, Kent, 1977), pp. 52–53.

12. See Carol Cosgrove Twitchett, "Towards a Community Development Policy," in *Europe and the World*, ed. Ken J. Twitchett (London: Europa, 1976), pp. 168–72.

13. *Commission of the European Communities*, "Memo," P-69, November 1975, p. 2. See also "Information Development Cooperation: Food Aid" (Brussels: Commission, 1977).

14. For example, see "Agricultural Expertise Exported," *Ireland Today* 910 (15 June 1977): 2–3 (Bulletin of Department of Foreign Affairs, Dublin).

15. "The Agricultural Policy of the European Community" (Brussels: Commission, 1976) no. 5, p. 3.

16. Ibid.

17. Ibid., p. 5.

18. Adrien Zeller, "European Agriculture in the World Economy," *A Nation Writ Large?*, ed. Max Kohnstamm and Wolfgang Hager, (London: Macmillan, 1973), p. 82.

19. These external issues are examined in John Marsh, "Europe's Agriculture: Reform of the CAP," *International Affairs* 53 (4 October 1977): 605. Marsh concentrates mainly on internal problems, but occasionally demonstrates the external effects of CAP's contradictions as well.

20. Ibid., p. 604, and especially Zeller, p. 90. In both cases conclusions were quickly out of date.

21. Cited in *Farm Feature* (European Communities Commission), Dublin, 29 November 1978, p. 1. See re milk gluts "Dairying Must Get Its House in Order," *Farmer's Digest* 42 (5 November 1978): 35–38.

22. Cited in *Farm Feature*, 27 September 1978, p. 2.

23. Ibid.

24. See "European Community Background Information" (EC Information Service, Washington, D.C.), no. 10, 30 May 1978, p. 1. See also, for effects of CAP, the *Herald Tribune* (Paris), 14 August 1978, p. 7.

25. For example, see Thomas B. Curtis and John R. Vastine, *The Kennedy Round and the Future of American Trade* (New York: Praeger Publishers, 1973), ch. 1–7.

26. See *The German Tribune* 866 (20 November 1978): 2.

27. One of the best examples of the dilemmas of GATT concerns Danish ham exports to the United States, which are deemed to be "subsidized." The value of these exports is $170 million and is about 40 percent of total Danish meat exports. This high-quality product sells for 20 to 25 percent *more* than U.S. products, and 10 to 15 percent more than the other main exporter's ham (Poland). The duties will add 8 percent to the price of Danish ham, based on the amount of subsidy the U.S. executive branch estimates the Danish farmer receives from the EC. See the *Los Angeles Times*, 18 December 1978, part 3, p. 14.

28. "Information: External Relations" (Brussels: Commission, 1978), p. 8.

29. See the *Los Angeles Times*, 19 December 1978, part 3, p. 10.

30. *Farm Journal* 102 (12 November 1978): 2.

31. Ibid., p. 28, and *Farm Journal* 102 (11 October 1978): 32.

32. "Information: External Relations," p. 11.

33. For just the oppsoite conclusion see John Karlik, "Economics and American Foreign Policy," in *The Interaction of Economics and Foreign Policy*, ed. Robert Bauer, pp. 31–32. EC and American positions at 1976 talks are outlined in the *New York Times*, 21 January 1978, finance, p. 37, and 12 July 1977, p. 39.

34. See *European Community* 205 (January-February 1978): 42; *European Community* 208 (July-August 1978): 40.

35. The proportion of total employment in agriculture is: Greece, 37.5 percent; Portugal, 31.1 percent; Spain, 28.6 percent. The highest proportion in a current EC member is Ireland, 26.5 percent. From "Basic Statistics," *OECD Economic Surveys*, Paris, March 1973.

36. See, for example, *The German Tribune* 825 (5 February 1978): 7; *The German Tribune* 827 (19 February 1978): 5.

37. *Bulletin of the European Communities* 3 (1978): 64–65. For one member state's perspective see *Developments in the European Communities*, Eleventh Report, January 1978, pp. 10–11 (Department of Foreign Affairs, Dublin).

38. For a discussion of the green international see Carey, op. cit., ch. 4. and 5, and David Balaam and Michael Carey, "The Influence of Domestic Constraints on Foreign Policies Regarding the Issue of Food: Comparison of the U.S. and the EC" (unpublished paper, March 1977), pp. 51–72.

39. Particular weaknesses that characterize French agricultural interest groups are analyzed in Balaam and Carey, pp. 62–65, and in William Averyt, *Agropolitics in the European Community* (New York: Praeger Publishers, 1977), ch. 3. See also Michael Leigh, "Mediterranean Agriculture and the Enlargement of the EEC," *World Today* 33, no. 6 (June 1977): 207–14. See also the *Los Angeles Times*, 13 September 1979, pt. 1-A, p. 5.

40. This is persuasively argued in Loukas Tsoukalis, "A Community of Twelve in Search of an Identity," *International Affairs* 54 (3 July 1978): 437–47.

41. Ibid., pp. 440–44.

42. Ibid., p. 444.

43. See Angelika Volle and William Wallace, "How Common a Fisheries Policy?", *World Today*, February 1977, pp. 67–72; D. C. Watt, "The EEC and Fishing: New Venture into Unknown Seas," *International Affairs* 48, no. 3 (July-September 1977): 328–36; *The German Tribune* 817 (11 December 1977): 6, 818 (18 December 1977): 2; and *The Irish Times*, 1 August 1977, p. 10, 20 August 1977, p. 1, and from one member state's perspective, *Estimates of the Minister for Foreign Affairs*, 1978 (Dublin), pp. 61–64.

44. "Mediterranean Agricultural Problems," *Newsletter on the CAP*, 5 May 1977, p. 1. (Brussels).

45. Ibid., p. 3.

46. Simon Harris and Dan Swinbank, "Price Fixing under the CAP—Proposition and Decision," *Food Policy* 3 (4 November 1978): 266.

47. *European Community Background Information*, p. 1.

48. See, for example, Wolfgang Hager, "The Community and the Mediterranean," in *A Nation Writ Large?*, pp. 195–221.

49. Ibid., p. 210.

50. Loukas Tsoukalis, "EEC and the Mediterranean: Is 'Global' Policy a Misnomer?" *International Affairs* (3 July 1977): 436–37. For an excellent critical analysis of how EPC towards the Mediterranean "fell" together see Avi Shlaim, "The Community and the Mediterranean Basis," in Twitchett, pp. 77–120.

51. Ibid., p. 118, Shlaim continues, "By acting in this way as a civilian power it will minimize the stress on the institutionally weak Community system, eschew friction with the superpowers and respond to the real expectations of its neighbors."

See also Hehmet Ali Berand, "Turkey and the European Community," *World Today*, February 1978, pp. 52–61; David Allen, "The Euro-Arab Dialogue," *Journal of Common Market Studies* 16 (4 June 1978): 323–42.

More general EC bilateral relations with the Third World are described in Timothy Josling, "Agricultural Trade Policies," *Round Table*, July 1977, pp. 277–83; *Information: Development Cooperation, Food Aid*, vol. 165, 1977, (Brussels), *The European Community and the Third World*, November 1977, (Brussels), "Information: Development Cooperation," (Brussels: Commission, January 1977).

4

Agri-Policy in the Soviet Union and Eastern Europe

DAVID N. BALAAM AND MICHAEL J. CAREY

Politics and fortune have combined to strain Soviet and East European agriculture and foreign trade policies. These conditions have contributed to a greater realization by national leaders of the role agriculture plays in national economic development and to the increased politicization of agricultural policymaking. A change in national leadership views, which formerly saw agriculture in terms of support for industrialization goals, reflects a new trend toward the adoption of policies designed to meet consumer demands. Increasing regional dependency on Western grain imports, conditioned by these domestic policies, paralleled a shift in the climate of international politics where relaxed tensions between East and West increased the salience of Soviet and East European agricultural policies at the global level.[1]

The relatively new position of the region as a regular grain importer is more acceptable to the Soviet Union than to Eastern Europe. The Soviet Union continues to import large amounts of expensive grain in the face of the American grain embargo. But the debt and dependency on international markets have caused many East European countries to renew efforts to attain agricultural self-sufficiency. However, self-sufficiency may ultimately be unattainable because of natural limitations and because development strategies continue to emphasize industrialization at the expense of agricultural growth. Self-sufficiency may also be irreconcilable with the continuing dependency on international markets to meet consumer demands for more and higher-quality foods.

The authors wish to thank Seth Thompson and Steve Newlin for their comments on earlier versions of this chapter, and Dan Pearson and Cindy Connally for their untiring manuscript assistance.

In order to understand the current politics of food in the Soviet Union and Eastern Europe this work assesses (1) the natural limitations that influence production and distribution in the region, (2) national food and agri-policies, and (3) regional trade patterns and organizational developments. Future world trade patterns are likely to be strongly influenced by this region's agricultural and trade policies, which are in turn influenced by the changing climate of international politics.

Soviet Agriculture

The debate about the success or failure of Soviet agriculture is an extremely controversial issue that will not be resolved here.[2] However, until 1971 the Soviet Union was a relatively self-sufficient food producer and a net grain exporter. Based on production successes in the late 1960s, beginning in the early 1970s Soviet leaders initiated a policy to improve the consumption of grain products and meat. This goal was part of an economic policy initiated by Brezhnev in 1964. However, during the 1970s Soviet agriculture was to experience regular periods of production shortages, in 1972, 1975, 1977, and 1979 (see Table 4.1). But the Brezhnev politburo did not revert back to a policy of self-sufficiency; instead, it shifted to Western food commodities, wheat, and feed-grain imports to meet consumer expectations and to develop Soviet livestock industries. Currently, the American grain embargo on the Soviet Union threatens the continuation of the new policy. If the 1980 grain crop results in another shortfall as serious as the one in 1979, livestock herds may be slaughtered earlier than planned, causing a significant decrease in per capita meat consumption.

There are many reasons why Soviet leaders have adopted and now continue to pursue a policy that makes them vulnerable to unpredictable domestic agricultural production and food supplied through international markets. This section focuses on some of those reasons related to the natural limits of Soviet agriculture, agricultural policymaking, and Soviet agriculture trade policies that have been greatly politicized over the last decade.

NATURAL LIMITATIONS

One factor that has played a large part in Soviet agriculture is growing conditions, which one Soviet agriculture expert has categorized as "handicapped by a colossal mismatch between soil and moisture."[3] The biggest problem is precipitation. Much of the Soviet Union is too dry to establish normal production patterns. Rainfall is quite variable, ranging anywhere from 8 to 23 inches a year. Only 1 percent of Soviet farmland receives 28 inches of rainfall a year, an amount that nearly all American states receive in one year. Droughts brought on by hot dry winds regularly occur every three years. Another problem is soil conditions. The *chernozen* (rich, black soil) of the steppes gets little rainfall. Where rainfall is plentiful the land is covered by coniferous forests and the soil is acidic. Still another natural barrier to production is the ground temperatures, which, because they are

Table 4.1 Soviet Area, Yield, and Production of Grain

	1966-70	1971	1972	1973	1974	1975	1976	1977	1978	1979
Area (1,000 hectares)	122,083	117,937	120,158	126,738	127,187	127,920	127,760	130,344	128,465	126,351
Yield (metric tons per hectare)	1.37	1.54	1.40	1.76	1.54	1.09	1.75	1.50	1.85	1.42[a]
Production (million metric tons)	167	181	168	222	196	140	224	196	237	179[a]

[a] Preliminary.

Source: USDA, "USSR Agricultural Situation: Review of 1979 and Outlook for 1980," and a variety of USDA publications.

50

extremely cold most of the year, cause short growing seasons and necessitate winter wheat and vegetable crops.

In many ways these conditions have shaped the agriculture policies of different Soviet regimes. For example, after a severe drought in 1946 Stalin attempted to divert the Black Sea into the Caspian in order to prevent the drying up of the latter. Stalin (and others) attempted to divert rivers in conjunction with efforts to expand the area of land under irrigation. Later, Khruschev concentrated on expanding the area of land under production. Brezhnev has tried to improve agricultural yields by land extension and land improvement programs that require added agri-inputs.

Until the early 1970s natural restrictions did not keep Soviet agriculture from achieving self-sufficiency. However, they are the confines within which Soviet agriculture now exists and which partially limit agriculture's ability to respond to the increasing consumer demands that result from population growth and increased per capita incomes. If only for reasons related to natural barriers, Soviet agriculture may have reached its limit of production capability. These factors certainly make policy planning difficult. Given declining labor supplies, they force leaders to count on added agri-inputs like fertilizer, machinery, and even capital to increase agricultural output.

AGRICULTURE POLITICS AND THE GREAT TRANSFORMATION

Current Soviet food and agriculture policies since 1917 have been conditioned not only by environmental factors but also by fundamental and systematic changes in leadership goals and decision-making styles. Before World War I Russia was considered to be the "breadbasket of Europe." Russia was one of the world's largest producers of flax, barley, cotton, and wheat, and exported on the average of 10mmts of grain a year.[4] Thus, Lenin's War Communism policies from 1918 to 1921 assigned to agriculture the role of supporting industry. Industry was to be financed by agricultural exports, which accounted for 40 percent of all exports. As part of a plan to centralize both bureaucratic authority and the direction of industry, peasant produce was confiscated by compulsory acquisition by detachments of armed workers and poorer peasants. Peasant producers responded by noncooperation, even violence, and withheld grain from markets.[5]

At the Tenth Party Congress in 1921 Lenin replaced his War Communism policies with the New Economic Policy (NEP), one goal of which was peasant-state (rural-urban) reconciliation, part of a larger plan to socialize the economy gradually when socialist enterprises could duplicate private enterprises.[6] Based on his assumption that capitalism and socialism could exist side by side, Lenin revised the role of market forces, restored private enterprise in smaller-scale industries, and introduced a more modest tax on peasants of 10 percent of output.

However, the NEP did not solve, but instead compounded, the state's "agrarian problem." NEP policies that still emphasized industrial development produced a "scissors crisis" whereby farmers were squeezed between high-priced industrial goods and low procurement prices held to the 1914 prewar level. Farmers could now legally hoard their commodities and wait for higher prices. In general, prices were largely incoherent and regionally inconsistent, acting as a "brake" on production. This forced the state to resort to the "gray market"—i.e., illegal middlemen, who provided as much as one-quarter of the grain needed to cover food shortages in urban areas. To compound the problem, land reform under Lenin had not gone as planned. Lenin wanted large estates to remain intact and to be run as state farms. But after the revolution, peasants quickly confiscated farms and hastily divided them up, which cut into large-estate and Kulak production trends. Before the war they had produced 70 percent of marketed grain. Furthermore, even though peasants participated in the earlier grain confiscation program, kulaks and poorer peasants did not develop opposing class interests, as Lenin had expected.[7] Under NEP policies Soviet agriculture remained divided into a variety of production units—only about a third of which were socialized— and some kulaks actually acquired more land.

Therefore, inconsistent procurement prices and land reform policies[8] that did not coordinate the host of production enterprises that dotted the countryside contributed to inefficiency and reduced agricultural output to prewar levels. The Soviet leadership, now under Stalin, suffered through two drastic production shortages in 1923–24 and 1927–28. Not only was the state's food supply for urban areas severely cut back, but the "resources"[9] extracted from agriculture that served as earnings for industrial investment were also severely threatened. The scene was set for the more drastic and comprehensive measures Stalin would adopt to deal with the problem.

Beginning in 1929 and ending in 1936 with 90 percent of Soviet agriculture socialized, Stalin pursued the goal of collectivization, at times in a ruthlessly brutal manner.[10] Perhaps best known as a device he used to purge and secure his regime from an "enemy within,"[11] collectivization served a number of other purposes. In terms of economic thought, collectivization did what capitalism could not do fast enough; the coexistence of capitalism and socialism under the NEP had not brought about the socialist economy Lenin believed it would. Collectivization served to rid agriculture of private enterprise in a straightforward manner. Farmers were allowed to maintain small private plots a little more than an acre in size, and to market their own intensively produced goods. But their relationship to the "means of produc-tion" was redefined. The main production unit became the one-quarter million collective farms (kolkhozy) which organized about seventy-five households on two square miles of land. Legally, collectives were coopera-tives and not state enterprises, a distinction that relieved the state of maintaining responsibility for wage and fringe benefits equal to those that urban workers received. Another 4,000 state farms (sovkhozy) were intended to be used as "grain factories" for experimental agricultural development projects.

The collectivization of the peasantry served to tighten state control over the food production and distribution process. Party officials directly super-vised Machine Tractor Stations (MTS) that supplied collectives with equip-ment, labor, and other agri-inputs in exchange for a percentage of produc-tion. The MTS also served as a means of Party control over social behavior.

They were charged with eliminating subversive elements who Stalin believed would sabotage agriculture. Stalin purged at least a million kulaks, who were either deported or relocated. Finally, by extending almost total state control over agriculture, the state planned to receive the lion's share of benefits derived from agriculture. In pursuit of "basic industrialization at any cost" the resources extracted from agriculture could better serve the purpose of rapid industrialization, and large-scale production would release more labor for the same purpose.

In terms of increasing agricultural output Stalin's efforts were not successful. Production actually stagnated. Between 1928 and 1953 (excluding 1940–48) annual production averaged only 0.9 percent per year; there was an 18 percent increase over the entire period. Livestock and grain production dropped. Collectivized agriculture resulted in the overextension of the state bureaucracy into agriculture and the loss of peasant incentive to increase production beyond subsistence levels. Peasants either hoarded food or consumed it before quotas were due.

Finally, collectivization contributed to the further separation of urban and rural populations. Without a doubt, the peasantry underwrote the "Great Transformation" of the Soviet Union, as the countryside bore a greater burden in terms of the sacrifice made to industrialization. Indeed, it is quite remarkable that even though certain peasant classes were eliminated—especially the kulaks, who were the most efficient producers—and price disincentives stood in their way, the peasantry managed to produce at self-sufficiency levels. In the face of such adversity, the Soviet Union achieved remarkable growth rates, not because but in spite of the exploitation of the peasants.[12]

In 1953, in an effort to de-Stalinize the economy and realize the fruits of a socialist economy in a more positive fashion, Khrushchev made his mark on Soviet agriculture. Facing an agricultural crisis, he took immediate steps to reform Stalin's policies and increase agricultural output. He raised state procurement prices, reduced compulsory sales to the state, cut taxes on private plots, and cut transport and marketing costs. In 1954 he also initiated the virgin lands program, which added some 74 million acres of land to production in southern Russia, Kazakhstan, and Siberia over a three-year period. This was a less costly approach to increasing production since land extension was substituted for more expensive inputs that were in short supply.

Perhaps Khrushchev is also best known for his general economic policies[13] that aimed at decentralization of the bureaucracy and at producing more goods consumers had been deprived of during Stalin's rule. At the Twentieth Party Congress in 1956 the Soviet leadership adopted a new goal: to move "Soviet agriculture from command farming according to physical plans dictated by the central authorities towards a market oriented agriculture guided by prices, stimulated by incentives rather than coercion, and based on rational specialization."[14]

Toward this end Khrushchev further relaxed restrictions on both public and private acreage. Peasants were allowed to increase the size of their private plots and livestock herds. Prices paid for compulsory quotas were increased, quotas were raised only when needed, and taxes were reduced.[15] On state farms, these policies had the effect of freeing capital for additional investment and establishing a guaranteed wage of sorts. At the highest levels of government Khrushchev attempted to decentralize the agricultural policy machinery by abolishing central ministries and the State Economic Commission, and replaced them with regional economic councils. This move actually strengthened Party influence over regional and provincial councils. In 1957 Khrushchev divided provincial and local party organizations between industry and agriculture in order to give more attention to agriculture. In 1958 he abolished the MTS, whose function as input supplier was not immediately replaced.

In 1958 these reforms reached their point of maximum return. To a great extent they successfully increased production, beginning in 1953 and culminating with a bumper crop year in 1958. The five-year period saw a 40 percent increase in total output, a 7.1 percent average annual production increase, a 4.4 percent increase in area sown, and a 0.6 percent increase in labor applied to agriculture. Against a background of initial production successes, then, and to further ideological goals, in 1958 Khrushchev rescinded some of his private subsidiary farming policies. The size of family plots was cut by 9 percent, and peasants were encouraged to sell their cattle to the kolkhozy. As part of a renewed effort to further collectivization, grain sales were cut to households and limited to collectives. A consequence of this was to further deter production. By 1958 it became clear that land extension would not replace needed capital and agri-industrial inputs. Some 40 million hectares of grasslands in Kazakhstan and the RSFSR were plowed up and sown to grain, which moved the Soviet Union's breadbasket region eastward nearer to the arid steppes. Bad farming practices and poor management resulted in a heavy loss of topsoil and moisture, causing many areas to become dustbowls during years of drought. Khrushchev also unsuccessfully attempted to restructure crop patterns, but most notably failed to make corn the main feed grain of Soviet agriculture. Thus, from 1959 to 1964 net farm output totaled only a 13 percent overall increase or a 1.3 percent annual increase. Sown area declined by 2.3 percent and labor engaged in agriculture decreased by 1.9 percent.

Not only did peasants have less incentive to produce but a number of Khrushchev's sweeping administrative reforms, including the reorganization and liquidation of Machine Tractor Stations in 1958 as well as the division of agriculture and industry at the provincial and local levels, failed because of administrative duplication and lack of direction. Procurement and pricing policies were not consistent betweeen or within different regions and between communal farms. Income restrictions did not permit farmers to cover rising production costs. As a result, collective farms found it difficult to adjust themselves to all the hastily conceived and ineptly implemented organizational and institutional changes Khrushchev instituted. The poor crop year of 1963, and the administration mismanagement that had occurred since 1958, played a major part in Khrushchev's removal from office in October of 1964.

Just as Khrushchev inherited the environmental and political problems of the previous regime, so Brezhnev inherited Khrushchev's problems, especially those related to a tendency for agriculture to operate well below its potential. However, Brezhnev moved more slowly to reform both agriculture and the general economy than did Khrushchev immediately after taking office. His agriculture goals and policy reforms mark a continuation in the effort to modernize Soviet agriculture, albeit with the objective in mind of agriculture "catching up" to industry via more cautiously designed programs that reflected consideration of changing political and economic realities. Actually, the resolution of intraparty debates during Khrushchev's last years set the stage for Brezhnev to adopt more market-oriented policies. At the 1962 November Plenum of the Communist Party, liberal forces (who based their economic aims on the views of economics professor E. Liberman) sought to decentralize the administrative decision-making structure and to rely less on planned targets and more on profit incentives. They won out over conservatives who "where simply not prepared to decentralize the allocation of capital and allow prices to move freely"[16] and instead argued for the increased use of computers to handle the centralization of work loads.

Therefore, shortly after taking office in 1964 Brezhnev adopted a number of temporary measures to enhance production incentives, including relaxed restrictions on private subsidiary farming and increases in several farm commodity prices.[17] But Brezhnev also overturned much of Khrushchev's ministerial and bureaucratic decentralization reorganization at the different government levels. Brezhnev's own outline of a broader economic and agriculture program did not begin to take shape until the March 1965 Central Committee meeting. Later at the September meeting of the Central Committee Kosygin made it clear that a lag had developed between industry and agriculture. At that meeting he stated that

The lag in agriculture . . . has created a certain discrepancy between the production of consumer goods and the production of the means of production of consumer goods and the production of the means of production. This could not fail to influence the rate of growth of the real incomes of the population and the standards of material incentives. The elimination of this lag by raising the efficiency of the whole of social production is the principal task today.[18]

Brezhnev's broad plan to decentralize the bureaucracy and provide for greater worker management, which would, he hoped, increase both incentives and the production of better quality consumer goods and thus stimulate economic growth, also included agriculture reforms. Procurement prices for all commodities were increased, with livestock given the biggest raise. Procurement quotas were lowered to more realistic levels and set in advance, and farmers were promised bonuses 50 percent above procurement prices for overplan deliveries. Government investment in agriculture was to increase dramatically from the 43 billion rubles spent on agriculture between 1961 and 1965 to a planned 71 billion between 1966 and 1970. Special attention was to be given to higher yields through irrigation and land drainage, machinery purchase programs, and increased fertilizer output. In 1966 Brezhnev also reformed the kolkhoz income tax base. Kolkhozniki (collective workers) were guaranteed monthly payments, with added prospects of bonuses for surpluses. Sovkhozy were also reformed in an attempt to make them self-reliant and to increase their decision-making autonomy. Finally, private plot and livestock ownership restrictions were again relaxed. In order to enhance efficiency, kolkhozy and sovkhozy were allowed to experiment with private subsidiary enterprises by making "full use of seasonal surpluses of labor and of materials available on the farms."[19]

These policies were particularly significant on a number of counts. Brezhnev's effort at increasing production incentives in a coherent, organized fashion was unmatched by none of his predecessors. These policies added to collective farm policymaking autonomy. Between 1965 and 1967 these policies also paid off in terms of increased agricultural output, especially in livestock, milk, eggs, and even grain, while the average annual production growth rate jumped to 4 percent. More importantly, agriculture was reformed more than industry—up to then the principal preoccupation of the leadership. Finally, Brezhnev succeeded in narrowing the income gap between collective farmers and urban workers. In fact, increased incomes since the early 1960s had a broader effect on the economy. As the Soviet population and per capita incomes grew (see Table 4.2), so did the demand for more and better-quality (i.e., protein) food, specifically meat and grains.

Since 1968 however, Soviet agriculture has not fared as well. Brezhnev's policies designed to meet these demands were not all that successful. While

Table 4.2   Soviet and East European Population and GNP Growth Rates

	Population (000) mid-1978	Population Growth rate 1970-77 (in percent)	GNP Per Capita 1978 (in $ US)	GNP Per Capita (real) Growth rate 1970-77 (in percent)
USSR	261,234	0.9	3,700	4.4
Poland	35,081	1.0	3,660	6.3
Yugoslavia	21,933	0.9	2,390	5.1
Rumania	21,853	0.9	1,750	9.9
GDR	16,859	−0.2	5,660	4.9
Czechoslovakia	15,120	0.7	4,720	4.3
Hungary	10,672	0.4	3,450	5.1
Bulgaria	8,892	0.6	3,200	5.7
Albania	2,606	2.5	740	4.1

Note: Excluding Yugoslavia, estimates of GNP per capita and its growth rate are tentative. For estimation procedures, see Technical Note, p. 22 in the *1979 World Bank Atlas.*

grain output increased dramatically during the first three years of his leadership, from 1967 to 1974 it fell off to only a 26 percent increase compared to a 41 percent increase under Khrushchev. The average annual increase in yield per hectare (refer to Table 4.1) was also to drop in the 1970s. It seems that techniques for increasing grain production by using added agri-inputs reached their point of marginal return about 1968.[20] Weather and a number of other factors, namely inefficient management, efforts to continue collectivization (albeit more slowly), external controls on farm decision making, and a return to some of Khrushchev's old land extension and amelioration programs after 1972, contributed to production setbacks. These conditions, combined with a series of droughts in 1972 and again in 1974, 1976, and 1979, led to serious crop shortages and forced the government to seek grain imports from the West.

Faced with the possibility of further dependence on these imports and vulnerability to international markets, the Soviet leadership has continued to pursue the goal of producing more consumer goods and increasing meat consumption. The current Soviet food policy seeks to provide Soviet citizens with an average of 165 pounds of meat per year by 1980. Yet agricultural production remains hostage to climate and soil conditions, and as per capita incomes grow, goods and services fall behind production forces that do not freely respond to market signals. State agricultural policymaking also remains hostage to the fear of a loss of centralized control over that process, which is why Brezhnev has been hesitant to officially support the more personal "link" (*zveno*) system of farming on collectives.[21]

However, the role of subsidiary farms in Soviet agriculture was publicly supported by Brezhnev at the 1976 Plenum and later in the Soviet press. It was hoped that these farms would contribute to the production of food and other light industry consumer goods, increase employment opportunities, and generally improve rural living conditions. However, the private sector's share of production has declined since 1970.[22] Private meat production was down from 35 percent of the total share in 1970 to 29 percent in 1980, while private milk and eggs were down from 36 and 53 percent to 29 and 34 percent respectively. Also in decline was the private-enterprise production of potatoes and vegetables. The main cause of this trend may be the many serious droughts that have occurred over the past decade. While the private sector has not overcome all of Soviet agriculture's structural and political problems, its utilization has resulted in providing significant supplies of food to collective and state farmers and to urban workers. Income earned from the private sector equals roughly 25 percent of the collective family farm income.

In conclusion, it may be said that Brezhnev's policies manifest a more coherent attempt to incorporate the agricultural sector—along with the Soviet consumer—into the general economy. Despite reform and changes, however, problems in Soviet agriculture remain. Current economic and agriculture policies represent an abandonment of food self-sufficiency goals. Production shortages have occurred more often, and agriculture has not been able to maintain the production consistency leaders have hoped for. How the leadership can continue its domestic food policies aimed at increasing the

Table 4.3 USSR Grain Exports To Selected Countries: Five-Year Averages 1956-1970, Annual 1971-1977, (in 1000 metric tons)

Destination	1956-60	1961-65	1966-70	1971	1972	1973	1974	1975	1976	1977[a]
Albania	68.1	10.7	–	–	–	–	–	–	–	–
Austria	67.0	65.0	35.0	91.7	–	–	–	–	–	–
Belgium	46.1	31.7	9.6	19.8	–	–	–	–	–	–
Denmark	28.0	14.3	–	–	–	–	–	–	–	–
Egypt	203.7	–	359.9	–	–	–	–	–	–	–
Finland	336.5	175.1	34.9	33.3	29.5	5.1	–	–	–	–
FRG	94.8	138.8	36.1	56.5	–	–	–	–	–	–
France	46.3	16.3	–	–	–	–	–	–	–	–
Italy	46.1	134.0	97.5	148.6	–	–	–	16.5	–	–
Netherlands	211.2	102.6	83.1	83.8	–	–	–	–	–	–
North Korea	38.2	98.2	170.3	152.2	156.6	181.8	181.1	212.2	212.6	292.1
Norway	120.2	70.4	20.5	–	–	–	–	–	–	–
Sweden	63.9	61.4	22.9	1.9	–	–	–	23.5	–	–
United Kingdom	103.5	306.5	195.8	282.9	–	–	–	–	–	–
CMEA										
Bulgaria	95.9	99.1	–	–	–	–	–	–	–	–
Czechoslovakia	1389.1	1239.4	1354.7	1489.2	1090.8	1094.1	668.7	587.1	56.4	312.6
GDR	1596.5	1559.0	1334.9	1912.7	1066.5	977.7	1425.1	719.8	187.2	873.4
Hungary	201.6	248.8	156.1	425.9	–	–	–	–	–	–
Poland	669.0	577.3	944.0	2132.6	1179.9	1069.2	1898.2	1014.6	268.8	795.9
Rumania	158.9	80.0	–	–	–	–	–	–	–	–
Yugoslavia	173.3	14.3	–	–	–	–	–	–	–	–
CMEA total[b]	4122.5	4224.5	4313.3	6534.3	3909.4	3749.4	4562.2	2874.9	1049.0	2834.0
Other[c]	13.0	91.8	109.7	830.2	453.8	830.0	2286.2	432.0	206.4	267.5
Total exports	5906.3	5879.8	5622.8	8639.6	4560.2	4853.3	7029.5	3578.0	1468.0	3393.6

a Calculated.
b Excluding Yugoslavia.
c Residual.
Source: USDA, "USSR Agricultural Trade 1955-77: "Foreign Agricultural Service."

production of consumer goods and more and better foods can best be understood in the context of Soviet trade policies.

SOVIET AGRICULTURAL TRADE

From 1955 to 1971 the Soviet Union exported an average of 3 million metric tons (mmt) of grain a year (except for the poor crop years between 1964 and 1966) mainly to Eastern Europe, Egypt, Cuba, North Korea, and Britain (see Table 4.3). But in the late 1960s and during the 1970s Soviet agricultural exports decreased, while agricultural imports increased significantly, both as a result of more frequent production shortages and changes in domestic food policies. By 1975 agricultural products made up 23 percent of all imports.[23] At the same time, Western grain exporters provided the Soviet Union with an opportunity to continue to pursue its new consumer-oriented policies at a lower cost than would have been possible if Western producers had not been saddled with large surpluses.

The trend of increasing agricultural imports occurred not only as a result of changes in domestic policy objectives, but also within a broader context of changing East-West relations during the 1960s and 1970s.[24] Soviet decision makers sought to reduce tensions with their old Cold War rival the United States. Expansion of Soviet trade with the West, especially with the United States, also evolved from these new goals of Soviet foreign policy during the 1960s and 1970s. Tensions with the PRC made it prudent for the Soviet Union to seek greater stability in relations with the West.

These foreign policy goals then, coincided with Soviet efforts to raise per capita consumption levels, improved East-West relations, and the opportunity to purchase low-cost grains. The Soviets turned to increased agricultural imports, specifically wheat and feed grains, to cover food shortages and develop Soviet livestock industries.

Regional and world trade patterns reflect this change in Soviet food policies and international market opportunities. In the 1960s the Soviet Union exported as much as 3.5 mmts of grain in one year to Eastern bloc countries, especially Czechoslovakia, the GDR, and Poland (refer to Table 4.3). As relations improved with the West and in accordance with changes in domestic policies, the Soviet Union exported less grain to its political allies and gradually imported more wheat and feed grains from Argentina, Australia, and Canada. American grain exports did not play a major role in Soviet trade until 1971 (see Table 4.4), after which time the Soviet Union purchased about two-thirds of its agricultural imports from the United States.

Western Europe and the Soviet Union together produce more grain than the United States, Canada, Australia, and Argentina. However, during the 1970s, the latter four accounted for almost 50 percent of world grain exports. Soviet decision makers turned to these countries most often for grain imports (see Table 4.5), although they continued to import from Hungary and Rumania. The Soviets regularly made purchases of Western grain in varying amounts, depending on domestic production and reserve supply levels and

Table 4.4   U.S. Agricultural Exports to USSR by Volume (Selected Grains):
Five-Year Averages 1956-1970, Annual 1971-1979 (in 1000 metric tons)

Commodity	1956-60	1961-65	1966-70	1971	1972	1973	1974	1975	1976	1977	1978	1979[a]
Wheat	—	337	—	3	2733	8727	1063	4118	1800	3017	2925	5400
Corn	1	—	—	471	3438	4816	2155	3433	9601	3766	10407	12000
Rye	—	—	—	—	—	682	160	—	—	—	—	—
Barley	—	—	—	—	816	106	—	—	98	—	—	—
Oats	—	—	—	41	306	—	—	46	19	—	—	—
Rice	—	11	—	—	—	—	—	22	52	88	16	22
Grain sorghum	—	—	—	—	—	1	1	—	2	—	—	—
Total grain[b]	1	348	—	515	7293	14332	3379	7619	11572	6871	13348	17422[c]

a Preliminary
b Includes transshipments.
c Excludes other grain categories and residuals.
Source: USDA, "USSR Agricultural Trade 1955-77 . . .," and "USSR Agricultural Situation: Review of 1979 and Outlook for 1980."

Table 4.5 USSR Grain Imports: Five-Year Averages 1956-1970, Annual 1971-1977 (in 1000 metric tons)

Origin	1956-66	1961-65	1966-70	1971	1972	1973	1974	1975	1976	1977[a]
Argentina	4.2	173.7	132.0	196.2	–	–	669.9	1764.4	1260.6	410.8
Australia	–	528.2	27.4	277.1	478.6	844.6	580.8	1186.4	1262.7	959.1
Brazil	–	–	–	50.4	–	–	287.9	749.2	535.3	62.3
Canada	222.2	2000.9	1812.0	1805.1	4681.0	4205.2	458.5	2718.3	3099.0	2353.4
Mexico	–	–	113.1	151.0	–	–	–	–	–	10.1
PRC	33.8	–	–	–	–	–	–	–	–	–
United States	–	356.9	–	–	7239.4	15370.7	4143.0	7141.0	11962.0	6265.0
Sweden	–	–	–	–	304.8	406.1	83.1	23.5	–	19.3
CMEA										
Hungary	–	–	–	–	99.9	543.9	137.1	854.3	420.6	228.8
Poland	8.0	–	–	–	–	–	–	–	–	–
Rumania	66.5	148.6	119.4	–	209.9	182.7	–	380.5	327.8	–
Yugoslavia	2.8	3.8	–	–	–	–	–	–	–	–
CMEA Total[b]	87.7	149.7	143.7	–	309.8	726.6	137.1	1234.8	748.4	222.8

[a] Calculated
[b] Excluding Yugoslavia

Source: USDA, "USSR Agricultural Trade 1955-77 . . . ," Foreign Agricultural Service.

international market prices. When production was up to expected levels, large purchases were made as a hedge against the next shortfall. Wheat, corn, and soybeans accounted for a higher percentage of Soviet grain imports, and as the Soviets built up their livestock herds, imports of wheat decreased in importance compared to feed grains (see Table 4.6). The Soviets purchased most of their corn from the United States and Argentina and most of their barley from Western Europe and Canada.

Even though constrained by limited foreign exchange reserves, in 1972 Soviet grain traders demonstrated remarkable business acumen by moving quickly and secretly to bargain with a number of large American grain companies to purchase some 19 million tons of grain from the United States and 10 million tons from other exporters. A number of American grain traders were privy to information about expected Soviet production shortfalls and bought up large amounts of subsidized grain to sell to the Soviets.[25] The Soviets were also given a loan of some $200 million by the Agriculture Department's Commodity Credit Corporation (CCC) to make the deal.[26] Following widespread criticism of the grain deal by a variety of American and West European interest groups and government officials, and after a series of inadequate measures by the Nixon and Ford administrations to halt the "great grain robbery," the Ford administration and the Soviets signed a five-year agreement that regulated Soviet purchases. The Soviets agreed to purchase at least six and up to eight mmts of wheat and corn annually with an option to buy more depending on American output levels.[27]

By October 1979 grain production in some regions of the Soviet Union was down by 25 percent from 1978 levels, or 48 mmts below the planned target. Large American supplies again expanded Soviet purchase opportunities. Prospects looked good for expanding grain markets in the Soviet Union. American farmers were advised that they could plant all they wanted and that they would still be eligible for subsidies.[28] In June the Soviets began contracting grain purchases with the United States and Australia. In October, President Carter offered the Soviets up to 25 mmts of grain.

However, because of the Soviet invasion of Afghanistan, on January 4, 1980, Carter reversed his policy and suspended, among other items, shipments of agriculture exports to the Soviet Union, as part of an effort "to demonstrate that such an act of aggression would result in serious and costly countermeasures."[29] Suspended were grain shipments above the eight mmt limit of the Moscow agreement. Of the 21.8 mmt contracted for purchase by the Soviets, 5.5 mmts were already shipped, leaving 2.5 mmts left to be exported under the five-year accord. Also suspended were contracts for oilseeds and oilseed products and livestock and poultry products. Shortly afterward, the United States also suspended phosphate exports, which the Soviet Union needs to manufacture fertilizers.

President Carter hoped to disrupt the Soviet five-year plan aimed at increasing per capita meat consumption. The USDA's Foreign Agricultural Service estimates that the embargo will create a 50 mmt shortfall in feed grains, or a 13 percent cutback in Soviet livestock feed imports. Stated

Table 4.6 USSR Wheat and Coarse Grain Imports: Five-Year Averages 1956-1970, Annual 1971-1977, (in 1000 metric tons)

Imports	1956-60	1961-6	1966-70	1971	1972	1973	1974	1975	1976	1977[a]
Wheat	246.7	3482.0	2577.0	2300.0	8100.0	15200.0	2706.7	9145.6	6686.0	6296.6
Net import[b]	−4244.9	9.1	−2104.2	−5316.6	4209.7	11007.1	−2555.3	6480.9	5878.0	4559.6
Barley	39.8	–	–	179.3	2600.0	1900.0	284.0	1001.0	2244.0	43.0
Corn	91.8	4.5	317.4	880.8	4059.1	5379.1	3440.3	5548.0	11376.0	3918.6
Rye	–	–	–	–	100.0	1300.0	691.0	–	–	19.2
Oats	5.7	–	–	139.3	600.0	100.0	9.0	214.0	332.0	54.4
Residual	0.8	11.4	22.6	0.6	40.9	20.9	–	0.4	–	–
Total coarse grains	138.2	15.9	339.9	1200.0	7400.0	8700.0	4424.3	6763.4	13952.0	4017.2
Net import	−1276.5	−2391.0	−651.6	177.0	6730.1	8039.6	2656.8	5850.1	13292.0	2360.6

a Calculated
b Negative equals export.
Source: USDA, "USSR Agricultural Trade 1955-77 ...," Foreign Agricultural Service.

otherwise, the embargo could result in a 4 percent drop from the 57 kilograms (125 pounds) of meat per year the Soviets now consume. This reduction could have domestic political repercussions for Soviet leaders.

The likelihood of the grain embargo's success and its implications for regional food politics will be discussed in the conclusion of this book. However, in conclusion of this discussion of Soviet agriculture, we can identify a number of trends that have emerged. All Soviet leaders have realized the extent to which agriculture plays an integral role in the development of Soviet industry and economy in general, yet some have taken agriculture for granted more than others. Significant agricultural policy changes that have occurred since the revolution reflect definite leadership goals and styles of decision making. For Lenin and Stalin, agriculture served two interrelated purposes: first, as support for the regime, and second, as the basis of rapid development of the industrial sector. Lenin was not as willing as Stalin (via the extensive collectivization program he initiated) to make agriculture pay too great a price to achieve industrialization. Agriculture served the same two purposes for Khrushchev. But based on Stalin's failure to achieve planned output levels, and as part of a general reappraisal of the economy, Khruschev's agriculture policies reflect an attempt to modernize the production process and shift attention to consumer demands. By 1953 the Soviet economy had gone through a "rural transformation": a significant shift had occurred in the numbers of people engaged in agriculture labor from the countryside to industrial employment in urban areas. More regime support came from urban areas, where increased per capita incomes and the changes in consumption habits that accompany an urban lifestyle required agricultural success. Therefore, Khrushchev's earlier policies attempted to enhance peasant production incentives and reduce the bottlenecks in supportive administrative and off-farm back-up systems. Khrushchev also attempted to slow the exodus of labor from agriculture. As demonstrated in the case of the virgin lands campaign, and given a small agricultural labor force, he also relied on "quick fixes" to increase production. However, Khrushchev experimented with new production methods and demonstrated to later leaders the potential for increasing production via a number of market elements. Even more significantly, he was willing to trade off defense expenditures and increase the state's budget for agriculture to three times over what it had been under Stalin.

Under Brezhnev, agricultural policymaking reflects less meshing with leadership politics and more concern with production output and efficiency. In carefully-thought-out ways, the agriculture policymaking process has been more decentralized, and probably has been carried to politically acceptable limits. Private farming now plays a greater and even essential role in Soviet agriculture. Yet vagaries in the weather and production problems have hurt Soviet policymakers just when they have tried to increase per capita grain and meat production and consumption.

Thus, the politburo has attempted to continue to improve the Soviet diet by importing needed feed grains, instead of slaughtering animals as was done previously when production shortfalls occurred. After 1960 Soviet leaders felt they could no longer impose such hardships on the general populace without political risk, and they turned to overstocked international markets for grain supplies. Until recently, Soviet agriculture imports also enhanced the trade connection between the Soviet Union and the West, partially satisfying Soviet domestic and foreign policy goals and helping to overcome the limits of Soviet agriculture that had pushed the Soviet Union into international markets in the first place. Whether or not the Soviet Union continues to seek more grain in international markets or reverts back to a goal of self-sufficiency is conditioned by natural limitations over which it has no control and other knotty problems including defense expenditures. However, if investment in agri-inputs has reached its level of marginal utility, then increased investment may not necessarily lead to greater production. The opportunity to buy low-priced grains from the West, an integral part of East-West détente, has until now relieved the Soviet Union of the need to shift more attention to agriculture and to the combination of factors (e.g., added labor) that might increase yields.

East European Food Systems

The eight nations of Eastern Europe differ greatly, not only in levels of economic development but also in terms of the role of agriculture in their economies. This section discusses the natural limits to modernizing and increasing agricultural production in Eastern Europe, a variety of political economic issues that surround agriculture in these countries, and the role of agricultural trade, especially as it pertains to both national development objectives and regional trade patterns.

After World War II, bloc and bilateral relations with the Soviet Union set the tone for the adoption of the Soviet model of agricultural collectivization and industrial development in Eastern Europe. Since the early 1950s, the Soviet model has been either rejected or gradually refined to fit a wide variety of national political and economic conditions. In some Eastern European nations the socialist model has led to increased production and improved food distribution, whereas in others the model has not produced the same results. On the basis of changing political and economic relations between the center and the periphery, a new—more autonomous—relationship has emerged. In almost experimental fashion the Soviet Union now watches Eastern Europe for favorable results in increased agricultural output, efficiency, and the like.

NATURAL LIMITS: PRODUCTION REVIEW

Since 1970 food production in Eastern Europe has increased, albeit at a relatively slow rate (see Table 4.7) given the effort made to increase production. Growth has been greatest in Rumania and Hungary and slowest in Poland. A host of factors contribute to this condition, but one that continues to plague the Eastern Europeans is growing conditions.

The structural foundations of Eastern European agriculture and food

Table 4.7 Soviet and East European Indices of Total and Per Capita Agricultural Production, 1970-1979

(1961-65 = 100)

	1970	1971	1972	1973	1974	1975	1976	1977	1978	1979
Total										
USSR	136	135	129	155	145	130	153	149	163	146
Bulgaria	126	130	143	136	127	138	148	136	143	154
Czechoslovakia	120	127	132	140	146	139	136	147	155	144
GDR	116	115	128	127	140	132	126	137	143	143
Hungary	116	134	143	147	157	155	151	166	172	163
Poland	115	111	122	130	131	128	132	122	133	129
Rumania	110	136	153	143	147	150	188	183	186	193
Yugoslavia	115	128	126	132	151	149	157	165	155	158
USSR and Eastern Europe	129	131	130	148	144	133	150	148	159	147
Per capita										
USSR	125	124	117	139	129	114	134	129	140	124
Bulgaria	119	122	134	127	118	127	136	124	130	139
Czechoslovakia	117	123	127	134	138	131	127	136	142	131
GDR	115	114	128	128	141	134	128	140	146	146
Hungary	113	130	138	142	150	148	143	157	162	152
Poland	108	104	112	119	119	115	117	107	116	111
Rumania	101	124	139	129	131	133	164	159	160	164
Yugoslavia	107	118	115	120	135	133	138	144	134	135
USSR and Eastern Europe	121	121	120	135	130	119	133	131	139	128

Source: USDA, "Indices of Agricultural and Food Production for Europe and the U.S.S.R.: Average 1961-65 and Annual 1970 through 1979," Economics, Statistics, and Cooperatives Service, Statistical Bulletin no. 635.

66

production match some of those in the Soviet Union in that they have an unpredictable effect on agricultural output. In the northern countries weather conditions are usually cold and wet, which often causes shortfalls in Poland's grain production. Heavy summer rains or a lack of sunlight result in excess amounts of moisture in harvested wheat.[30] Rumania, Bulgaria, Czechoslovakia, and Hungary are susceptible to droughts and Yugoslavia to storms that damage corn, fruit, and vegetable crops. Overall, soil conditions in Eastern Europe are generally not as bad as they are in the Soviet Union, and some of the southeastern countries have longer growing seasons.

These conditions have made agricultural surpluses difficult to accumulate. Because of structural limitations, Eastern Europe's food self-sufficiency goals have been difficult to realize. As in the Soviet Union, the increased demand for food that results from increased income earnings and population growth rates that have gradually slowed since 1970 (refer to Table 4.2) has necessitated grain imports from the West.

AGRICULTURE AND REFORM POLITICS

After World War II, Eastern Europe's agricultural economies were significantly transformed by the attempted implementation of the Soviet Union's agricultural-industrial development model. Rapid industrialization was the primary goal. Rural sector subsistence farming was to be modernized by a collectivized agricultural sector that would absorb economic shocks, feed the urban-industrial sectors with food and labor, and use industry's output to mechanize production. Socialist planners linked farm production to industrial production in an effort to transform the peasant culture into a rural industrial organization.

The Eastern Europeans attempted land reform and promoted collectivized agriculture on the Soviet model.[31] But in many countries collectivization was judged inappropriate to local conditions and was met by resistance. Yugoslavia abandoned the goal as early as 1952. Poland did likewise in 1956.

Eastern Europe's collectivization goals were slowed significantly in the late 1950s and 1960s, although by no means were they completely abandoned. In Yugoslavia and Poland nearly 85 percent of agricultural land is now held in private ownership. In other countries as little as 10 percent of the land has gone back to private ownership or subsidiary farming. Although land reform was carried out more successfully in some countries than in others, until later it did not significantly change the subsistence nature of agriculture in many East European countries. Eastern European agriculture is still socialized agriculture, but resistance to collectivization has been greater in Eastern Europe than in the Soviet Union, not only because of its inappropriateness to indigenous conditions but also in opposition to Soviet hegemony.

In conjunction with the problems of socializing agriculture, many Eastern European countries realized that the industrialization goals of the Soviet "command economy" model would be unattainable given the different structures of East European economies.[32] For example, Czechoslavakia and East Germany had industrial labor shortages, while Poland and Hungary had an oversupply of agricultural labor. Most of the economies were also dependent on foreign trade, which proved to be irreconcilable with industrialization goals. The highly centralized administrative economic model proved especially detrimental to the agriculture sectors of East Germany, Poland, Czechoslavakia, and Hungary, where unregulated procurement policies, unrealistic targets, and unfavorable pricing schemes led to declines in national productivity and labor inefficiency. The result was agricultural production shortages, land use and crop distortions, manpower shortages, food gaps that imports could not cover, and low farm incomes. Many countries were left without exportable products with which to earn income needed for trade.

The death of Stalin made way for reform of the Soviet model. Khruschev's de-Stalinization campaign laid the groundwork for the adoption of more realistic goals and decentralization of national agricultural policymaking. Yugoslavia began the process by developing its own model of worker enterprise management and administrative decentralization. The other economies began to follow shortly thereafter.

Eastern European agriculture reforms[33] can be divided into three interrelated categories: (1) changes in general economic thought and resultant agriculture policies, (2) administrative reform, and (3) attempts to increase farm incomes. By the early 1960s most Eastern Europeans preferred a less centralized "intensive" development plan to the old "extensive" highly centralized variety. Many reformers were impressed by Western economic performance and the reformist ideas of Liberman. Emphasis shifted from rapid structural change to more realistic expectations about output increases, development goals, and the introduction of markets within the boundaries of a looser centralized planning scheme. The political significance of this effort was that reformers could "argue the need for change without as such condemning the traditional system, with which the Communist Party leadership was so intimately involved."[34]

The case of administrative reform at the national level was much like that of the Soviet Union, in that it came often and was inconsistent. Rumania, East Germany, Bulgaria, and Czechoslovakia shifted agricultural policy decision making away from the Ministry of Agriculture and either merged it with broader agencies like the Council for Agriculture and Food Industry or, as in the case of Hungary and Czechoslovakia, allowed for the establishment of Central Cooperative Unions. Compulsory targets were significantly decreased in almost all nations. Higher government administrative officials were to still decide "main proportions," while lower level associations or directorates were to decide production plans, investment allocation, and the like. Enterprises were given more autonomy. To stimulate production, farmers were allowed to increase their subsidiary activities and were given more bargaining power to deal directly with procurement agencies. By the late 1950s low-priced compulsory deliveries were eliminated in some countries and single price systems were established in Hungary, Rumania, Bulgaria, and Czechoslovakia. In some countries (Czechoslovakia and Hungary) tax systems were also reformed to allow farmers to cover production costs.

In terms of farm incomes, these reforms resulted in significant income increases, although large country wide disparities still exist. Private plots were given a greater role in Hungary, Bulgaria, and Yugoslavia. The private sector now accounts for one-third of Hungarian and Rumanian agriculture and about one-fourth of production in Czechoslovakia and Bulgaria.

Albania has seen the smallest amount of reform, while Hungary, with its earlier "guided market" and "new economic mechanism," and Yugoslavia have realized the most. The reforms in Poland, East Germany, and Bulgaria have been of limited consequence, while until recently, Czechoslovakia's reform efforts were stalled by the aftermath of the Prague Spring. Despite both general economic and agricultural reform, newer agricultural policies in Eastern Europe have not lead to substantial net production gains. But some countries have fared far better than others. The cases of Poland[35] and Hungary[36] are good examples of nations that dealt with the problem in different ways. Poland's growth rates have been some of the slowest in Eastern Europe. Ninety percent of farmland is privately owned, but the remaining socialist sector produces almost 20 percent of total agricultural output. In Hungary, 88 percent of farmland is collectivized and accounts for 65 percent of production, earning Hungary the distinction of being "champion of production" in Eastern Europe.[37]

After World War II, Polish leaders resisted collectivization but did emphasize industrial growth. Not only was Poland heavily dependent on Soviet industrial imports, but between 1950 and 1970 agriculture's share of total investment decreased significantly from 32.5 percent to 15.2 percent. When the economy "overheated," increases in per capita incomes and a jump in population led to increased food demands. The government responded by using up reserves and finally raising consumer food prices, which incited food riots in 1970. Gierek replaced Gomulka and froze food prices for two years. Production continued to lag, and as it did, the government was forced to absorb tremendous subsidy costs. In 1975 the government tried to slow consumer consumption and increase agricultural production. In 1976 consumer prices were raised once more, producing the same political result.

Since then, Polish agriculture has become caught in a tighter squeeze. Even though one-third of the work force is still in agriculture (see Table 4.8), its numbers have declined at a fast pace. The typical farm continues to be a "dwarf" farm, under five hectares. City life and higher wages attract the young. Those who remain are the elderly who cannot work as hard and either lack the money to invest or resist using more expensive agri-inputs.[38] Because of the debt already incurred by the government for subsidies, it resists investing more capital in agriculture. Instead, it has attempted to keep the young "down on the farm" by a series of measures designed to increase the size of farms including pensions to older farmers for release of uncultivated or neglected land and loans to private farmers to purchase or rent

Table 4.8   Soviet and East European Division of Labor Force (in percentage)

	Agriculture		Industry		Services	
	1960	1977	1960	1977	1960	1977
USSR	42	19	29	46	29	35
Poland	48	34	29	38	23	28
Yugoslavia	64	42	23	34	13	24
Rumania	64	51	20	31	16	18
German Dem. Rep.	18	10	48	51	34	39
Czechoslovakia	26	13	46	49	28	38
Hungary	38	19	35	58	27	23
Bulgaria	57	41	25	38	18	21
Albania	71	63	17	24	12	13

Source: The World Bank, *World Development Report* 1979, August 1979, p. 163.

farms. The government has also abolished compulsory deliveries and now contracts prices and advance payments. Cooperative ventures in agribusiness are officially supported, along with higher price supports.

Polish farmers have not been anxious to accept these measures. They complain that pensions are too small and still prefer to sell their produce in private markets. Meanwhile, agriculture production has started to rise again. But food shortages attributed to poor weather conditions have to be made up by imports (discussed below), which does not help Poland's balance-of-payment situation.

There are some similarities between Poland's and Hungary's agriculture sectors. Hungary's development strategy reduced agriculture's share of national income from 47 percent in 1938 to 18.7 percent in 1971. As in Poland, there has been a dramatic shift of the labor force out of agriculture. In 1977 only 19 percent of the total labor force was engaged in agricultural work (refer to Table 4.8). As in Poland, the Hungarians resisted collectivization and abolished collectives in 1953. Also as in Poland, agricultural production in Hungary increased very slowly until the 1960s.

However, unlike Poland, Hungary began to pursue collectivization again in the early 1960s. This time resistance was significantly diffused by a strategy that targeted all classes of peasants to receive benefits. The state accepted more decentralized agri-policy decision making at the local level and the continuation of private farms and household plots, which more efficiently produced labor-intensive products like small animals, fruit, and eggs. Finally, the government supported the Technologically Operated Production Systems (TOPS),[39] an imported Western specialized operation unit that efficiently increases output of a variety of crops or manages production processes. TOPS are legal entities whose shareholders receive profit dividends. TOPS corn producers doubled corn acreage and output by the use of hybrids. The result of these developments was a remarkable turnaround in agricultural production rates in Hungary, which has been exporting its agricultural model to other East European countries and to LDCs.

Bulgaria, Yugoslavia, the GDR, and Rumania have initiated reforms similar to those in Hungary in order to increase yields.[40] In all Eastern European countries efforts to increase production have been expensive for governments in terms of the cost of agricultural investments and subsidies. In the 1970s production costs increased, largely because of more expensive raw material and technological imports. Governments were still hesitant to raise consumer prices, while fixed producer prices that rose slowly still outran consumer costs and forced the state to pay large subsidies.[41] Because of inflation and a general economic slowdown in industrial sectors,[42] much of Eastern Europe's agricultural labor force has not been able to relocate to cities. Thus, due to the added cost of urban programs, capital investment in agriculture is expected to remain a problem.[43]

EAST EUROPEAN TRADE PATTERNS AND POLICIES

In 1963 the United States ended its commercial trade embargo on the East.[44] Grain had previously been shipped to Eastern Europe by the Western Europeans and others who did not uphold the embargo, and also by the United States as part of the P.L. 480 Food for Peace program.[45] Beginning in the mid-1960s, then, the West made a concerted effort to sell grain to Eastern Europe, which was provided with an opportunity to turn away from the Soviet Union for much of its grain supplies. After 1971 Soviet exports of grain to Eastern Europe (refer to Table 4.3) were to drop off sharply anyway when the Soviet Union experienced one of its worst production years and shifted emphasis to increased grain and meat consumption. Soviet grain exports to Bulgaria, Rumania, and Yugoslavia had already been cut in the mid-1960s, while Rumania continued to supply the Soviet Union with substantial amounts of grain.

During the 1970s Western grain exporters have competed for the Eastern European business of roughly 12–13 mmts of grain needed per year.[46] Eastern Europe's biggest importers are Czechoslovakia, the GDR, and Poland. Under the terms of a bilateral agreement Poland annually purchases 800,000 tons of grain from France and some 1–1.5 mmts from Canada. The EC has been a preferred supplier of grain to Eastern Europe because of proximity, which cuts transportation costs and grain prices that were until recently held low by EC subsidies. The EC has temporarily suspended these subsidy payments and has sold little grain to Eastern Europe in recent years. The vacuum has been filled by the United States, which has steadily increased exports to the region (see Table 4.9). In 1979 the United States accounted for 60 percent of the region's grain imports, or about 8.9 mmts of grain. As the Eastern Europeans have tried to build up their livestock herds their coarse grain imports have increased in volume over wheat. The United

Table 4.9 Volume of U.S. Agricultural Exports, by Selected Commodities, to Eastern Europe, (in 1,000 mmts)

	1971	1972	1973	1974	1975	1976	1977	1978	1979[a]
Wheat									
Bulgaria	–	–	–	–	–	–	–	–	–
Czechoslovakia	–	–	74	–	9	143	–	–	442
GDR	–	146	418	11	335	719	84	219	196
Hungary	–	–	–	–	–	–	–	–	–
Poland	334	142	837	199	502	698	637	584	817
Rumania	382	29	–	–	86	427	171	–	81
Yugoslavia	–	396	268	146	–	–	–	–	406
Total	716	713	1597	356	932	1987	892	803	1942
Feed grains									
Bulgaria	–	–	–	64	115	246	3	226	42
Czechoslovakia	259	98	46	5	–	769	81	398	810
GDR	403	556	742	1164	1626	2158	1248	925	1702
Hungary	–	–	24	–	–	–	112	106	1
Poland	459	306	908	697	1471	2101	1496	2063	2449
Rumania	32	183	121	512	534	239	242	327	898
Yugoslavia	316	420	–	52	–	–	–	269	1092
Total	1469	1563	1841	2494	3746	5513	3182	4314	6994
Total grains[b]									
Bulgaria	–	–	–	64	115	246	3	226	42
Czechoslovakia	259	98	120	5	10	912	81	398	1252
GDR	403	702	1160	1175	1961	2877	1332	1145	1898
Hungary	–	–	24	–	–	–	112	107	1
Poland	459	448	1745	896	1973	2799	2133	2683	3266
Rumania	366	212	121	512	620	666	413	327	979
Yugoslavia	698	816	268	198	–	–	–	269	1498
Total	2185	2276	3438	2850	4679	7500	4074	5155	8936

[a] Preliminary
[b] Includes transhipments

Source: USDA, "Eastern Europe Agricultural Situation: Review of 1978 and Outlook for 1979" and "Review of 1979 and Outlook for 1980," *Economics, Statistics, and Cooperatives Service.*

States also provides 90 percent of the region's corn imports, while Poland purchases two-fifths of American soybean exports.

The terms of trade for Eastern Europe have gradually deteriorated, and as world prices (especially for technology and raw materials) have increased since 1973, Eastern Europe has been saddled with credit and balance of payments problems. With the exception of Bulgaria, Rumania, and Hungary, the other nations run large trade deficits. Regional debt totaled $10.6 billion in 1977, $10.8 in 1978, and $11.3 in 1979, 20 percent of which was for agricultural products. Eastern Europe has received credit from a number of sources. Canada, France, and Great Britain have extended credit to Poland, the region's biggest agricultural importer. Funds have been provided by Western and CMEA banks,[47] bonds issued in international money markets (e.g., Eurodollars), through joint ventures with foreign firms, and commodity trade programs. The United States has been a regular supplier of credit to Poland, Rumania, Hungary, and Yugoslavia. Hungary and Rumania have been extended MFN status, along with Poland and Yugoslavia. The United States has also tried to promote agricultural trade through the 1978 Agricultural Trade Expansion Act by extending CCC intermediate credit to promote animal breeding, agricultural marketing, and the purchase of wheat for reserves. The USDA has also opened a number of trading offices in Eastern Europe.[48]

The amount of debt related to increased trade has forced many Eastern European food importers to reevaluate their national development strategies. Efforts have been renewed to attain grain self-sufficiency and to rebuild livestock herds. Many would like to decrease their imports of agricultural goods, but not necessarily the technological products and inputs used to increase yields. Yet, because of production shortfalls and increased consumer demand, trade patterns between Eastern Europe and its neighbors still reflect an increase of grain imports.

In conclusion of this section it is clear that agricultural production in Eastern Europe has continually been below planned target levels because, except in a few cases, governments have not invested enough capital in agriculture (following the Soviet model) or have been reluctant to raise consumer food prices. Thus, many Eastern European countries have been largely dependent on agricultural imports to cover food shortages and shortfalls, especially of the type that occurred in 1963–64. In many ways imports have been a heavier economic burden on Eastern Europe than they have on the Soviet Union, adding to the problem of generating economic growth and investing in agriculture for the sake of self-sufficiency. Agricultural imports therefore play a crucial role in Eastern European food systems and are likely to be politicized even further due to issues that create a tighter linkage between domestic conditions, trade policies, and general economic conditions.

Agricultural Regionalism and International Trade

Eastern Europe and the Soviet Union usually do not act as a region in international trade and politics. An economic regional organization, modeled on Western Europe's integrative efforts—the Council for Mutual Economic Assistance (COMECON)—was created in 1947 as an attempt to lay the foundation for economic integration.[49] COMECON has expanded its tasks and institutions (for example, the International Bank for Economic Cooperation was recently created), but it has not integrated Eastern European economies nor the Soviet–Eastern European policy process.

COMECON's initial progress was the result of Soviet support, the Cold War atmosphere, and the impact of the Western embargo that promoted interregional trade. Integration slowed during the mid-1960s when Western trade barriers came down and many Eastern European nations sought to increase trade with the West. Détente, too, led to a decline in COMECON integration because almost all the East Europeans prefer autonomy and bilateral trade.

This is the case in agri-trade. In Western Europe, increasingly, bilateral deals between the EC and individual Eastern European nations are the order of the day. The United States as well has recently concluded new agri-trade pacts with Bulgaria and the GDR. Although not a COMECON member, Yugoslavia has been a preferred customer of the West. COMECON would not deal with the EC because, until recently, the Soviets denied the authority of the EC; likewise the EC did not recognize the COMECON. This aspect of the East-West relationship has nearly been resolved; yet bilateralism on the part of Eastern Europe remains the dominant mode of interaction.

One of COMECON's biggest problems is currency convertibility. Eastern European currencies are not directly convertible; through bilateral negotiations, exchange rates are set, often for different types of products or activities.[50] Less than half of COMECON trade is with the Soviet Union.[51] Commodities with world markets are sold for exchangeable currencies, even to COMECON partners. Up to 10 percent of COMECON trade is also settled in U.S. dollars. Some specialty food items are sold to the West to finance imports, often high technology from the West. One of the most regular trade relationships, although a hidden one, is that of East Germany and West Germany, a link that may well be a channel for other Western trade. Another is the long-standing United States–Poland aid and trade relationship.

At the regional level COMECON and EC talks are stalled over the question of mutual recognition. The EC sees COMECON as a loose association that does not have the authority to negotiate or even to influence member tariffs. Yet Eastern Europe and the EC (and of course individual Western nations) continue to ratify bilateral trade pacts and joint commissions and endeavor to promote trade. This continues despite the Soviet invasion of Afghanistan because East European decision makers and those in the West do not want Soviet behavior to interfere with inter-European trade. Agri-trade deals with Rumania and Yugoslavia were recently concluded with the EC.[52] Barter transactions in agricultural products (and raw materials) will also probably continue to be important. Marketing bartered agricultural goods may not raise as many problems for traders as might be expected.[53] Financing agri-trade with convertible currencies is probably the most desirable method of payment, but it is one that presents difficulties for Eastern Europeans, who lack adequate supplies of hard currencies.

Eastern Europe, additionally, may help the Soviet Union overcome the grain embargo. East German imports through West Germany (the port of Hamburg) increased tremendously in the first months of 1980.[54] Much of the increase was grain, which, along with other items is likely to be transshipped to the Soviets. But the United States has accounted for this in its estimates of Soviet imports. Thus, while COMECON may not be a truly regional agri-trading body, on a bilateral basis agri-trade has developed to the benefit of Eastern Europe and probably to that of the Soviet Union.

Conclusion

Alterations in the politics of food in the Soviet Union and Eastern Europe since the early 1970s have led to significant changes in agricultural policies and regional trade patterns. Consequently, agriculture now plays a greater role in national economic strategies. New agricultural policies are likely to be politicized further because both natural conditions and political barriers continue to limit agricultural output in the region. Goals of increasing the production of meat and higher-quality food have required drastic increases of grain imports, most of which have come from the West.

The extent to which the Soviet leadership, except for Stalin, adopted policies that allowed for private subsidiary farming or market forces depended on food production successes and the availability of surpluses. Administrative and market restrictions were relaxed until a large enough production cushion could enable leaders to return to focusing attention on the collectivization of agriculture and on ridding the economy of private subsidiary farming. While this same trend holds true for some of the Eastern European countries after World War II, many of them resisted collectivization and reformed the Soviet model of development to fit indigenous conditions. Reform varied from country to country depending on leadership values, agricultural structural conditions, offsetting interest-group influences, and changes in Soviet leadership attitudes about hegemony over Eastern Europe. Many of the East European countries have been more willing than the Soviet Union to let market forces and private subsidiary farming influence food production and the price of producer and consumer goods.

Agricultural policies over the entire region have been significantly politicized because of an attempt by governments to meet consumer demands for increased grain and meat consumption, which have risen in response to growing populations and per capita incomes. This change in domestic policy coincides with changes in the climate of international politics, where a relaxation of tension between East and West—détente—and favorable world trade conditions provided the region with the opportunity to purchase grains from Western exporters who were seeking new markets. The Eastern Europeans took advantage of this opportunity to cover shortfalls with Western grain imports earlier than the Soviet Union, which did not abandon its goal of self-sufficiency until 1972. A shift in Soviet domestic policies that led to a slowdown in grain exports, along with the Eastern European preference for importing Western grains, contributed to a significant shift in regional and world trade patterns.

Because national policy differences remain substantial, this region has not developed agricultural integration to any great extent, nor is it likely to, given the tendency of nations to act independently of one another rather than in unison when dealing with outsiders. Little in the way of impetus toward regionalism exists besides Soviet hegemony, and that alone does not suffice to create a fully integrated agricultural region.

A focus on the food systems and agricultural policies of this region indicates the dilemmas within the region. At the national level is the issue of investment in agriculture versus other priorities. It would be easy to suggest that solutions to food problems in this region would be solved by substantial investments in agriculture. While this might be appropriate for some Eastern European countries and the Soviet Union, investment alone is not the solution to the problem. The question arises why some nations in the region have been rather successful agricultural producers, and is there any relationship between production success and socialized agricultural production units and processes? Comparison of the agricultural sectors in Poland and Hungary demonstrates that the type of agricultural system does not directly determine the success of the system. Poland's agricultural sector might benefit from larger economies of scale that more efficiently utilize agriinputs, can increase production faster, and are more labor-efficient given the exodus of farm labor to cities. Attempts to attract labor to rural areas may also be more efficient than adding only more capital and agri-inputs to the production process, given the high cost of imported agri-inputs since 1973.

Efficiency, then, is another matter that must be considered in relation to production. Efficiency is a matter of the relationship of cost factors to units of production, which vary greatly in different nations. What is most important is "the general desire and willingness of the regime to make significant investments in the agriculture sphere."[55] The issue remains whether the Soviet and other Eastern European leaders will be able to reconcile the desires for centralization and control over the production process with needed decentralized measures that in many cases would increase output and improve efficiency. Of course, leadership attitudes about the value of the peasantry or farmers, which do show signs of having changed significantly since World War II, also play a major role in determining production successes.

Hungary has made a greater commitment to agriculture via a number of interrelated measures to increase production than has Poland or the Soviet Union. Hungary has done more to integrate the peasant into the policy-making process, adding to regime support for collectivization efforts. The Hungarian government also supports private subsidiary farming of products that are more efficiently produced by the private sector and has interfered less in the pricing system. Recently it allowed consumer prices for basic foodstuffs to increase by 4.6 percent in order to allow market forces to set prices and control supply and demand. Apparently the Hungarian government is more willing to accept the political consequences of such a move.

In terms of world trade patterns, agricultural conditions in the region have important implications. While some countries may achieve self-sufficient production levels and do so very efficiently, other countries may not be able to achieve self-sufficiently, given natural limitations and national political and economic factors that compel them to import more than they are able to produce themselves. The Soviet Union, for one, has already come too far to willingly reverse its dependency on grain imports without great political risk. Faced with American grain embargo, it has turned elsewhere—to Argentina and Brazil—for its grain supplies. The consequences of this behavior may backfire on the United States, as occurred in 1975 when the United States embargoed soybean exports to Japan. Ultimately, the biggest cost to the world is the extreme politicization of grain trading by overt attempts to use grain as a tool of foreign policy, which might add to a further deterioration in East-West relations.

Soviet needs and preference to continue imports as opposed to reversion to self-sufficiency strategies has important implications for regional agricultural developments. Although it is too early to tell for certain, the possibility that the current embargo or any further attempt to cut off Soviet grain supplies could force the Soviet Union to revert to a strategy of agricultural self-sufficiency makes it even more necessary that the Soviet Union accept Eastern European autonomy, especially as it relates to Eastern Europe's attempt to be self-sufficient and thus less dependent on the Soviet Union for grain imports. At the same time, Eastern Europe serves as an experiment with a variety of strategies to increase production that the Soviets can watch. Eastern Europe may be the new model of agricultural development for the Soviet Union.

Notes

1. An earlier tendency of analysts was to neglect the Soviet Union and Eastern Europe's agricultural policies and policymaking. This may have been due to the Soviet position as a self-sufficient food producer and the minimum level of trade between communist regimes and the West.

2. For a more detailed discussion of the question of Soviet agricultural success see Harry Shaffer, ed., *Soviet Agriculture: An Assessment of Its Contribution to Economic Development* (New York: Praeger Publishers, 1977).

3. M. Gardner Clark, "Soviet Agriculture Policy," in Shaffer, *Soviet Agriculture*, p. 4.

4. Robert Osborn, *The Evolution of Soviet Politics* (Homewood, Ill.: Dorsey Press, 1974). Much of the discussion of Lenin and Stalin's agriculture policies are based on Osborn's work.

5. T. H. Rigby, *Lenin's Government: Sovnarkon 1917–1922*, (New York: Cambridge University Press, 1979).

6. For a good discussion of the political and economic dimensions of Lenin's NEP see Paul Scheffer, "The Crisis of the 'N.E.P.' in Soviet Russia," *Foreign Affairs* 7, no. 2, (January 1929): 234–41.

7. The reason that peasants and Kulaks did not develop opposing class interests is related to closer peasant relations with the Kulaks, who, despite their greater wealth, identified more with the peasants—their laborers—than they did with the urban workers, who wanted lower price supports for agricultural production. For a more detailed discussion of this problem related to land reform and to the "agrarian question" see Karl Kautsky, *The Dictatorship of the Proletariat* (Ann Arbor, Michigan: University of Michigan Press, 1971), especially pp. 101–19.

8. For a more detailed discussion of the politics of collectivization during this period see M. Lewin, "The Immediate Background of Soviet Collectivization," *Soviet Studies* 17, no. 2, (October 1965): 162–97.

9. We have purposely not used the word "surpluses" here, as is commonly done in discussions about agriculture serving the purpose of industrial development. Instead we refer to extracted "resources." This distinction is discussed by James Millar and Alec Nove in a series of polemical articles on whether agricultural surpluses underwrote Stalin's industrialization plan. Millar distinguishes between a variety of surplus products, only one of which was excess food supplies and argues that too often the industrial sector is seen as a parasite on agriculture (as opposed to focusing on the interdependencies between agriculture and industry). Thus, he can evaluate the appropriateness of Stalin's program. Nove considers Millar's distinction semantic and instead prefers to focus on leadership intentions. See James Millar, "Soviet Rapid Development and Agriculture Surplus Hypothesis" and Alec Nove, "The Agriculture Surplus Hypothesis: Comment on Millar's Article," *Soviet Studies*, vol. 22, no. 3 (July 1970): 77–93, 394–401.

10. For more detailed descriptions and a discussion of Soviet collectivization see Mikhail Sholokhov, *Seeds of Tomorrow*, (New York: Alfred A. Knopf, 1959) and Merle Fainsod, *Smolensk under Soviet Rule*, (New York: Vintage Books, 1963), especially chapter 12.

11. The words "enemy within" refer not only to Stalin's own personal enemies but also to the Marxist notion that peasants were largely a conservative, antirevolutionary social stratum that Lenin had not properly managed and that Stalin felt, in the form of kulaks among other peasant classes, might subvert agriculture and thus the economy.

12. Millar, "Soviet Rapid Development and Agriculture Surplus Hypothesis."

13. For a more detailed discussion of agriculture politics during Khrushchev's tenure, see the classic work by Sidney I. Ploss, *Conflict and Decision Making in Soviet Russia*, (Princeton, N.J.: Princeton University Press, 1965).

14. Morris Bornstein, "The Soviet Debate on Agricultural Price and Procurement Reforms," *Soviet Studies* 21, no. 1 (July 1969): 2.

15. For a more detailed discussion of government policies in the area of private plots and procurements during Khrushchev's term see C. A. Knox Lovell, "The Role of Private Subsidiary Farming during the Soviet Seven-Year Plan, 1959–65," *Soviet Studies* 28, no. 1, (March 1969): 46–65.

16. Marshall Goldman, "Economic Controversy in the Soviet Union," *Foreign Affairs* 41, no. 3 (April 1963): 506.

17. For two excellent discussions of policies adopted by Brezhnev after taking office, and the political economic environment in which they were implemented, see Jerzy Karcz, "The New Soviet Agricultural Programme," *Soviet Studies* 27, no. 2 (October 1965): 129–61, and Roger Clarke, "Soviet Agricultural Reforms since Khrushchev," *Soviet Studies* 20, no. 2 (October 1968): 159–78.

18. A. N. Kosygin, "Russia's New Five Year Plan," *Vital Speeches* 32, no. 1 (15 October 1965): 117.

19. Clarke, "Soviet Agricultural Reforms," p. 166.

20. Roy D. Laird and Betty A. Laird, "The Widening Gap and Prospects for 1980 and 1990," in *The Future of Agriculture in the Soviet Union and Eastern Europe*, ed. Roy D. Laird, Joseph Hajda, and Betty A. Laird (Boulder, Colo: Westview Press, 1977), pp. 27–47.

21. For a more detailed discussion of the link type of production unit see Dimitry Pospielovsky, "The 'Link' System in Soviet Agriculture," *Soviet Studies* 21, no. 4 (April 1970): 411–35.

22. USDA, "USSR Agricultural Situation: Review of 1979 and Outlook for 1980," *Economics, Statistics, and Cooperatives Service*, WAS-21.

23. David Schoonover, "Soviet Agriculture in the 1976–80 Plan" in Laird et al., *The Future of Agriculture in the Soviet Union and Eastern Europe*.

24. See Daniel Yergin, "Politics and Soviet-American Trade: The Three Questions," *Foreign Affairs* 55, no. 3 (April 1977): 517–38.

25. There is also evidence that the USDA had prior knowledge of Russian grain needs, and that on a visit to the Soviet Union before the deal was made, Secretary of Agriculture Earl Butz learned of their situation. This information was passed on to the grain companies. For a detailed

description of the 1972 Soviet Wheat Deal see William Robbins, *The American Food Scandal* (New York: William Morrow & Co., 1974).

26. It should be noted that the Soviets have tried to accumulate more hard currency by developing export markets in many Western countries for a variety of products including furs and cotton. At times they have also liquidated some of their gold reserves.

27. A similar example of Soviet business acumen is the "butter deals" with the European Community. The Soviets purchased large quantities of subsidized surplus butter. The chagrined Europeans have recently announced a system of export licenses to keep closer watch on butter sales to foreign countries.

28. American farmers strongly supported the Soviet "pipeline purchases" because of their guaranteed market features. Although they went along with the current grain embargo, for reasons of "national loyalty," they now question it because of the economic losses they have incurred.

29. USDA, *Statement by the Honorable Bob Bergland, Secretary of Agriculture, To the Joint Economic Committee, United States Congress*, USDA reprint 253-80, 30 January 1980, p. 1.

30. USDA, "Eastern Europe Agricultural Situation: Review of 1978 and Outlook for 1979," *Economics, Statistics, and Cooperatives Service*, WAS 18, Supplement 3. For a more detailed study of geographic conditions in Eastern Europe, see R. H. Osborne, *East-Central Europe: An Introductory Geography*, (New York: Praeger Publishers, 1967).

31. Parts of southeastern Europe had a basis for communal agriculture, left over from the previous Turkish administration. The northern countries had a different foundation; large-scale *latifundia* was predominant in Poland, East Germany, Czechoslovakia, and Hungary. For a more detailed discussion of Eastern Europe's rural-sector transformation after World War II see George Hoffmann, "Rural Transformation in Eastern Europe since World War II," in *The Process of Rural Transformation: Eastern Europe, Latin America and Australia*, ed. (New York: Pergamon Press, 1980), pp. 21–41.

32. For a more detailed discussion of Eastern Europe's attempts to reform the Soviet model pursuant to a variety of national structures see Jerzy F. Karcz, "Agricultural Reform in Eastern Europe" in *Plan and Market: Economic Reform in Eastern Europe*, ed. Morris Bornstein (New Haven, Conn.: Yale University Press, 1973), pp. 207–43. For the special case of Yugoslavia, see George Hoffman and Fred W. Neal, *Yugoslavia and the New Communism* (New York: Twentieth Century, 1962).

33. Ibid.

34. Morris Bornstein, "Introduction" in his *Plan and Market*, ibid., p. 5.

35. For discussions of Poland's agriculture, a source of many of the figures found here unless otherwise noted, see Jaroslaw A. Piekalkiewicz, "Kulakization of Polish Agriculture," in *The Political Economy of Collectivized Agriculture: A Comparative Study of Communist and Non-Communist Systems* ed. Ronald Francisco, Betty Laird, and Roy Laird (New York: Pergamon Press, 1979), pp. 86–107; and Jacek I. Romanowsky, "Prospects for the Future of Polish Agriculture; and Andrzej Brzeski, "Poland's Uncertain Five-Year Plan," in Laird et al. *The Future of Agriculture in the Soviet Union and Eastern Europe*, pp. 97–148.

36. For discussions of Hungary's agriculture and a source of statistics used here, unless otherwise noted, see Ivan Volgyes, "Modernization, Collectivization, Production, and Legitimacy: Agricultural Development in Rural Hungary," in Francisco et al., *The Political Economy of Collectivized Agriculture*, pp. 108–29; and Peter Elek, "Hungary's New Agricultural Revolution and its Promise for the Fifth Five-Year Plan," in Laird et al., *The Future of Agriculture in the Soviet Union and Eastern Europe*, pp. 171–84.

37. Karl-Eugen Wadëkin, "The Place of Agriculture in the European Communist Economies," *Soviet Studies* 29, no. 2 (April 1977): 238–54.

38. *New York Times*, 6 June 1980, p. 2.

39. For a more detailed discussion of TOPS in Hungary see Peter Elek, "Hungary's New Agricultural Revolution."

40. See Wadëkin, "The Place of Agriculture."

41. USDA, "Eastern Europe, Agricultural Situation: Review of 1979 and Outlook for 1980," *Economics, Statistics, and Cooperatives Sciences*, WAS-21, Supplement 3.

42. *New York Times*, 1 November 1979, part 1, p. 2.

43. Wadëkin, "The Place of Agriculture."

44. For an excellent discussion of the American trade embargo on the East during the Cold War years see Trudy Huskamp Peterson, "Sales, Surpluses, and the Soviets: A Study in Political Economy," in Richard Fraenkel, Don Hadwiger, and William Browne, *The Role of U.S. Agriculture in Foreign Policy* (New York: Praeger Publishers, 1979), pp. 56–79.

45. For further discussion see Vaclav E. Mares, "U.S. Aid to East Europe" *Current History* 51, no. 299 (July 1960): 36–44.

46. The USDA publication, "Eastern Europe, Agricultural Situation, Reviews and Outlook," 1978–79 and 1979–80, unless otherwise noted, are sources of statistics regarding current domestic agriculture and agricultural trade in Eastern Europe.

47. Recently, a joint venture brought together the National Bank of Hungary, six European banks, and a Japanese financial institution to form the Central European International Bank Ltd.

48. The USDA regularly publishes information regarding trade with both the Soviet Union and Eastern Europe. See, for example, "Trading with the USSR and Eastern Europe," *Foreign Agricultural Service*, FASM-264, June 1975.

49. For a review of the history of COMECON and the conditions upon which it was founded, see Andrzej Korbanski, *International Conciliation*, no. 549 (September 1964).

50. USDA, "Eastern Europe Agricultural Situation, Review of 1978, Outlook for 1979."

51. For a survey of current COMECON developments see Nora Bellof, "Comecon Blues," *Foreign Policy*, no. 31 (Summer 1978), pp. 195–79., Z. M. Fallenbuchl "Comecom" *Problems of Communism* 22 (March-April 1973): 25–39; and Nish Jamgotch, "Alliance Management in Eastern Europe: The New Type of International Relations," *World Politics* 27, no. 3 (April 1975): 409–29.

52. *Europe*, no. 219 (May-June 1980), p. 51.

53. Peter Hermes, "Foreign Policy and Economic Interests," *Aussenpolitik*, no. 29 (March 1978), p. 243.

54. *The Los Angeles Times*, 12 May 1980, part 4, p. 1.

55. Volgyes, "Modernization, Collectivization, Production, and Legitimacy," p. 118.

Part II
The Developing Regions

Regionalism, Food Policy, and Domestic Political Economy: Lessons from Latin America

RICHARD GANZEL

Those familiar with the Latin American tradition of state activism in economic matters may be tempted to dismiss much of the literature on international food policy and on related regionalist approaches as primitive efforts to reinvent the wheel. A more helpful response may be to acknowledge that the Latin American experience is too rich in insight to remain the terrain of regional specialists. Moreover, since a satisfactory approach to food policy ultimately rests upon a theory of political economy, the fact that the region is a focal point for many who are assessing the merits of competing paradigms in political economy makes knowledge of the Latin American experience with activist strategies essential for food policy specialists.

This essay attempts to provide both a critique of key assumptions and concepts drawn from the literature of food policy and related commentary on international aid and trade policy, and a more positive set of suggestions of what approaches and concepts may be more likely to clarify and resolve food problems. It argues that fundamental flaws limit the utility of both the "needs" and the "dependency" approaches to political economy, despite their respective insights and contributions. It advocates following the realist schools of international relations scholarship, as well as classical socialists, in structuring inquiry around an assumption that the state remains the primary

Thanks for research assistance from John Morrison and for helpful comments from Lynn Scarlett and Richard Moore.

decision maker responsible for the welfare of its citizens in both actual and ethical senses. It further advocates reviving the central historical insight of classical capitalist theorists—that societal wealth can be augmented only within limits defined by the degree of adherence to comparative advantages in economic endeavors and by the enlargement of the market that negotiations by their political leaders open to producers. If accepted, the logic of regionalism as a tool of the state is clear. OPEC-style organizations, like military alliances, are formed to assert or redistribute power capacity among states; if successful, they redistribute present wealth however defined. Trade and investment institutions negotiated to ensure mutual access to wider markets and investment pools are tools of specialization based upon comparative advantage, creating greater total wealth. Particular regional institutions, consequently, can usefully be evaluated and categorized in these state-oriented terms, rather than whether they are social, economic, or political.

The first part of the essay sketches a critique of the food policy literature in general, with particular attention to the Latin American experience. The second part concentrates on the experiences of Argentina, Brazil, and Chile, countries in many respects more economically advanced and diversified than many of their neighbors, and each with a distinctly different yet valuable agricultural potential. More importantly, each possesses features that have been singled out for attention by analysts of food policy and politics: (1) lengthy experience as a participant in the international system of production, trade, and investment centered in the North Atlantic region; (2) a tradition of concentrated land ownership that has been reflected not only in a dual society but also in a complex and evolving tension between status reflecting productivity and status primarily attributable to wealth; (3) an income distribution sufficiently unequal that a significant portion of the population has been undernourished; (4) a tradition of protectionism, in trade relationships, especially during the twentieth century; (5) heavy reliance upon "customs," including levies on exports, to supply important portions of public revenues (thereby magnifying the domestic consequences of commodity price swings but also revealing an administrative capacity to utilize revenues from trade in primary commodities to redirect economic activity); and (6) national experiences with more than one broad strategy intended to shape the course of social, political, and economic evolution.

A Critical Commentary

Much of the literature devoted to what Hopkins and Puchala have called the "Global Political Economy of Food"[1] rests upon assumptions and relies upon concepts that need to be rethought. Without implying that all scholars share the tendencies criticized, or engaging in an extended literature survey, it seems apparent that writers have placed heavy reliance upon aggregate, especially macroeconomic, approaches. They have either shared the judgment that national self-sufficiency (at minimum, in food production) is an appropriate policy target or accepted the conceptual constraints resulting from the assumption that such a goal will in fact guide national policies. They have assigned much of the ethical responsibility for "solving" the global food problem (and frequently also for causing it) to the affluent countries of the world. Finally, they view the market mechanism (sometimes translated, almost literally, as high food prices) as a key part of the food problem. Taken together, application of these assumptions and judgments leads inexorably to a strategy of international redistribution within which regional groupings can perform important functions as power blocs (e.g., OPEC) to enforce transfers of wealth or as larger frameworks of self-sufficiency operating at arm's length form and negotiating with developed centers in the capitalist and communist worlds.

The proponents of aggregate analysis must contend with a growing consensus that overall levels of development or present rates of economic growth may tell little about either the basic health of an economy or its condition with respect to food production and distribution. Chilean prosperity based on silver, copper, and nitrate revenues,[2] and the Brazilian monarchy resting on a succession of tropical products,[3] are important cases in point. The same hesitation should be extended to sectoral performances in manufacturing or agriculture, and to such aggregate indicators as rates of inflation. What one needs to know of any particular performances is how it has been achieved, whether it can be sustained without subsidy or suppression, at whose expense it has been attained, whether the losers are being reintegrated into the society, and so forth. Each of these questions requires a level of analysis substantially more restricted than that of national aggregates. Each question, moreover, requires analysis of dynamic interrelationships rather than data and projections that "sum up" the present and future state of those relationships.

The tendency to equate high food prices with "the food problem" undoubtedly is a manifestation of the concern for the needs of the destitutue that motivates many an attempt to understand the food problem. But such concern cannot grow food. It neither motivates landowners to strive for productive efficiency nor inspires subsistence farmers or peasants to produce a surplus beyond the needs of family and relatives. On the other hand, high prices and profit incentives are essential preconditions for the willing transfer of capital and effort from other endeavors into the agricultural sector. No doubt there may be other preconditions to such a transfer, such as access to land, or trained labor, or water rights, that must be met for high prices to result in expanded food production. In too many cases, those trained in microeconomics (which assumes the existence not only of competition but, more importantly, of cultural values suggested by the notion of economic rationality) have omitted this crucial qualification. Those who have studied the Latin American experience would insist that it be made explicit, not as a disclaimer of responsibility for an "imperfect" world not populated by economic man, but instead incorporated into basic concepts. For example, a recent text by Gary W. Wynia that follows in the analytical tradition of political economist Charles W. Anderson[4] distinguishes conceptually between *latifundistas* and commercial farmers.[5] The former, by definition, are oriented primarily toward social and political values; the latter, toward economic incentives. In many cases, each is a large landowner.

The commitment by the political economist to examine the role accorded to price incentives in a particular society, and to examine as well the operative values of landowners, by design leaves open the question of whether concentrated landownership contributes to the food problem. Highly concentrated ownership patterns in Latin America historically have not in general been associated with land use decisions oriented toward maximizing production or vigorously applying scientific innovations to agriculture. But there have been important exceptions, individually and for extended periods in several countries, leaving causal relationships obscure. In Soviet society, a dictatorship pledged to egalitarian distribution has had unimpressive results with a program of land reform based on collectivization and socialization. In contrast, and despite Jeffersonian rhetoric that attributes democratic virtue to the family farm, maximization of production and scientific innovation indisputably have gone hand in hand with the trend toward concentrated land ownership in the United States. In certain Asian countries, differing soil, rainfall, and climatic conditions have yielded highly efficient productive units on a much smaller scale producing quite different foods.

These examples demonstrate that questions of economic efficiency must be separated analytically from those addressing cultural norms or the distribution of rewards within societies. At the same time, analysts must remain conscious of the historical evolution of these factors—or more precisely, their evolving interrelationship—in particular countries. The following quotation drawn from the conclusion to Ernest Feder's study of Latin American agriculture illustrates the danger of failing to separate these dimensions:

[L]atifundismo as a whole is an agriculture of unemployment, and we must not expect the forced increase in the use of science and technology to improve sharply overall performance of the agricultural sector. We must not expect that aggregate agricultural output, as opposed to the output of a few specialized crops, will be raised in any spectacular manner. . . . It is more than likely that the remaining latifundio sector, in continuation of past trends, will not participate in the modernization process and will turn, as in the past, increasingly to extensive land uses. . . . Increasingly extensive land uses are the best mechanism to continue monopolized ownership of land. . . . This does not necessarily imply less monetary returns. . . .[6]

Clearly, in many circumstances, economic incentives fail to alter land use practices making modification of land tenure patterns and/or other reforms necessary. Also, large landowners may "waste" land resources by employing them less than optimally in grazing. But small farmers may deplete soil by practicing slash-and-burn agriculture, hasten its erosion by ignoring elevation contours, and otherwise practice undesirable husbandry of a crucial resource. Feder is wrong in implying that grazing land is inferior to cultivating it. Optimal use depends both on the nature of the resource and its location. It also is difficult to make sense of distinctions between output of a few specialized crops and agricultural output in the aggregate. And how does one avoid reduced monetary returns if one uses land for grazing rather than cultivation, unless grazing is the optimal use? Most importantly, the implication running through this passage and the book it summarizes is that large numbers should be employed in agriculture to improve distribution *and* production.[7]

Except in a limited and transitory sense not intended by Feder, the opposite more nearly is true. Neither low agricultural productivity nor inappropriate land use will be solved until large numbers of people have been displaced from rural areas. For just as unproductive large landowners are a problem, so the problem with a peasantry is that it *is* a peasantry. Its situation will not be relieved except temporarily by putting its families in control of two or five acres instead of only one. Worse, the ready availability of unskilled labor for temporary employment at menial wages, reflecting both low productivity and labor surplus, is a crucial condition for long-term viability of a *latifundia* system.

Thus, insistence that land holdings be on a scale suitable for application of scientific technological and managerial techniques frequently is the institutional concomitant of an informed societal decision to devote agricultural resources to uses dictated by comparative advantage and the market. A particular country may not have land resources of sufficient quality for its products to compete in international markets. But almost all will have lands that, because of proximity to national markets, can successfully compete with imports to supply a portion of the people's food needs. For national as for international markets, specialization and division of labor is important. Productivity must be maximized, just as so-called marginals must be mobilized to produce goods that can be exchanged for needed imports.

There is no doubt that high food prices will exacerbate the hardship faced by those displaced from the countryside, even as they create a demand for labor in the "truck farming" belts surrounding urban areas and provide an incentive for outlying landowners to participate more vigorously in the national market. Neither the need for massive income transfers nor the need for vastly expanded educational and training opportunities for displaced peasants is likely to be met satisfactorily in practice. Consequently, there is ample justification for strategies designed to slow rates of urban migration during an extended period of transition. But the policymaker who loses sight of the goal of an agricultural sector dedicated to efficiency, rather than self-sufficiency for a large agrarian population, will have made the problem worse.

Similarly, the widely shared belief among Western analysts and policymakers that it is the obligation of affluent societies to solve the "food problem" of the poor in less fortunate societies must be rethought. When it comes as a call for restitution from those who have labored under colonialism, such an assignment of ethical responsibility is at least understandable. Similarly, analysts who maintain that organizational patterns of trade that have resulted in part from international legal, financial, and trade institutions, have distributed benefits unequally among developed and less developed states can construct a case for restitution, though the case is difficult to demonstrate. As Robert W. Tucker has forcefully argued, the claim upon more affluent societies appears to rest upon an extension of egalitarian values, in the process ignoring political and economic realities in affluent and poor countries alike.[8]

There is first the reality that many in affluent countries are not affluent. Does the obligation also apply to them, and if so, why? Second, for many (most?) poor countries, powerful affluent elites dominate. As a succession of scholars, most recently and from an egalitarian perspective Richard R. Fagen,[9] have reminded us, most international redistributive mechanisms perforce must work with and through such elites. In such circumstances, it is easy to reach quite different assessments of obligation. But in any case, it would be distinctly surprising if international transfers would intentionally be permitted to do more than provide minimums essential for the precarious subsistence of the poor. Denied Malthusian sanctions by international subsistence transfers, and denied fundamental reforms by domestic elites, the world's population of unproductive poor is thereby permitted to increase. Ironically, the unintended consequence of such "humanistic" aid programs may be the perpetuation and reinforcement of national systems of domination.

The international "dilemma" constituting the conditions within which ethical responsibility must be assessed and assigned is not strikingly different from that which Norman K. Nicholson and John D. Esseks have sketched for India:

The moment one accepts the inevitability of securing a substantial portion of its emergency reserves domestically, the problem of price levels presents itself. Those affected by hunger have little earning capacity and cannot pay market prices for food. A subsidy is the obvious answer. [But] given the difficulty of raising taxes in third world countries, the transfer payments which relief expenditure represents are not usually transfers from the rich to the poor but from the poor to the poor. Every rupee of relief could have been used toward productive investment elsewhere.

. . . If the government decides to buy in the market, however, the presence of such large "commercial" purchases will tend to force prices up, to the advantage of the better capitalized farmer who can wait to take advantage of the unusually high prices which prevail after the government has drained the market.[10]

Unfortunately, Nicholson and Esseks fail to surmount the logic of the short run. For them, private rationality is an evil to be overcome; giving in to it permits unearned benefits for the prosperous out of the hardship of the poor. The easy answer of public subsidy remains the obvious answer. The dilemma, in fact, is the hard choice of sacrificing in the present—lives via starvation in many cases where population growth has been unchecked too long—to devote scarce resources to investments. Only national regimes can make such choices. If they choose investment, outsiders can augment their investments. In fact, useful investments typically will be made through mutual interest. But if national regimes choose present subsidy, outsiders can only expand the number that are subsidized and who must suffer or change course later. Needless to say, except for emphasis, the course of subsidy fails to create incentives for private investors to join forces with governments in the struggle to end poverty by raising productivity.[11]

The situation is tragic rather than a true dilemma. The "obvious solution" of subsidy and aid is no solution at all. The good it does today is more than offset by the harm that follows tomorrow. Biologist Garrett Hardin's controversial argument in behalf of "lifeboat ethics" may be more insightful,[12] expecially if one recognizes that the so-called tragedy of the commons results from an absence of adequate control to ensure that resources are devoted to uses that can be sustained permanently. Thus reformulated, optimal solutions for particular resources could range from private ownership to global management.[13] Following Hardin, any solution that sacrifices long-term needs for present needs must be regarded as unethical, however pure the motives upon which it rests.

Latin America Food Policy

The effort to compare development paths of countries that have differing resources, cultures, and consumption patterns is extraordinarily difficult. Judgments drawn from such comparisons must be limited and tentative. A key problem is the absence of common standards of monetary valuation based on consumption and investment patterns within the different societies being compared. Instead, most economic comparisons rely upon the exchange values of national currencies. These values are heavily biased toward items that enter international trade. Moreover, comparisons in particular years may be distorted because exchange values are especially sensitive to temporary international capital flows and capital flights that respond to a variety of factors.

Despite such qualms, the comparative ranking of Latin America as a region and of its individual countries provides a general perspective that is useful. According to a recent World Bank study,[14] only Haiti ranks among low-income countries. Only Columbia, among Latin American countries with more than ten million people, has a per capita income below the world mean average of U.S. $750. The 1976 estimates offered for the three countries whose experience is examined more closely in the second part of this essay are: Chile, $1,050; Brazil, $1,104; and Argentina, $1,550.

Turning from income to food comparisons, Hopkins and Puchala note that the Latin American region is an overall net importer of cereals, though the deficit is not large.[15] Looking more closely, James E. Austin suggests that the levels of caloric consumption for 36 percent of the people are inadequate, with diets of 23 percent of the people substantially below minimum daily requirements.[16] These figures reflect the impact of income distribution on food consumption in the region. If the poor had adequate incomes, either the food trade deficit would grow or domestic production would be stimulated to fill the gap. Domestic trends show significant promise. Per capita food production is estimated at 4 percent above levels of a decade ago, despite soaring populations.[17] The performance of Argentina, an important exporter of grains to Latin American and world markets, has matched average gains for the region. Brazil, the most populous country and a net importer, produces 14 percent more food per capita than it did a decade earlier.[18]

Although such aggregate data is suggestive, it tells us little about how these production totals and recent trends have been achieved, nor what levels of production might exist if different strategies had been followed. A decade ago, Joseph Grunwald and Philip Musgrove took a different approach.[19] Using the best available information on natural resource endowments, they trace the experiences of Latin American countries with extracting, processing, and utilizing the proceeds derived from those resources. As a work in applied theory, their study emphasizes ways in which the national returns from primary products traded internationally have been affected by governmental policies and elaborates the industrialization sequences followed both in terms of governmental policies and of resource bases.

The hope is that, by turning from slippery numbers to a solid base of known resources, one can develop estimates of whether particular factors of production such as energy and minerals are sufficiently abundant to become "cheap" inputs for an economic sector dependent upon them—in short, whether they yield a comparative advantage in a real sense. Some of the analysis provided remains useful today, as does the informed speculation of Grunwald and Musgrove on what might have been if different governmental policies had been adopted. But their international comparisons of advantage were marred by reliance upon exchange values. One must still grapple with the differing production and consumption patterns in separate nations, because comparative advantages always depend upon the most efficient *combinations* of production factors and transportation costs that can be achieved. Just because X can be produced more cheaply in one society than another doesn't mean it should be. The advantage may be even greater in specializing in production of Y. Moreover, numbers attached to resources are no less subject to slippage than are exchange values. New discoveries regularly alter these numbers and so do technological breakthroughs. In theory, then, resource values are even less firm than are currency exchange values.[20] In practice, however, major discoveries or breakthroughs are more rare, giving substantial stability to resource-based trend analyses.

A recent study by Jorge Salazar-Carillo tackles the relationship between external and internal values, or prices and purchasing power comparisons, for a number of Latin American countries.[21] It is not possible to summarize underlying assumptions or comment critically upon the complex statistical analyses here, since the scale and importance of the study warrants a full-length evaluative article to do justice to it. However, if one accepts the estimates provided by Salazar-Carillo (and uses present technology and resource base data), it is possible to estimate international comparative advantages within the Latin American region. Of course, outside countries might have even greater advantages that offset transportation costs. Resource bases and technology might change sufficiently to upset these calculations. Finally, numerical precision undoubtedly is misleading, such that small advantages must be regarded as undemonstrated, if for no other reason than the notorious unreliability of statistics for the region.

Salazar-Carrillo provides tables showing whether a particular product type (food) is cheap or expensive compared to other product types (consumer durables) in particular countries. He then constructs a regional index comparing the relative cost of particular product packages in different countries, weighted for respective consumption patterns. Consumers, especially the poor, are vitally interested in whether they must pay more for staples they prefer than poor neighbors must pay for their preferences. These tables are also valuable to those analysts who focus upon the ability of the poor in various countries to meet basic needs in culturally acceptable ways, rather than through caloric abstractions.

Domestic price levels and price relationships (beans/rice) inevitably reflect existing patterns of price controls, exchange rate, credit, budgetary, and trade policies, and may bear little relationship to underlying economic realities. Nevertheless, Salazar-Carrillo labels these indices "absolute comparative advantage," and attempts to estimate from them how much efficiency for the region might be improved if trade restrictions were removed but governmental policies toward the internal economy were left intact. The regional specialist will recognize that formula as the minimalist rationale for the Latin American Free Trade Area that has won few committed adherents. Stated in reverse, no member state would be required to change domestic policies, including policies toward exchange rates, but manipulation is not assumed.

What overall pattern would result from removal of trade barriers within a Free Trade Area regime? Salazar-Carrillo concludes:

the smaller and less developed Latin American countries, with the exception of Bolivia and the addition of Colombia, appeared to be potential exporters of consumer products. The larger and more developed countries, Argentina excepted, seemed to be well suited to the potential exportation of investment products. Argentina had some export possibilities in the line of non-durable consumer goods, while Venezuela apparently could potentially exploit certain lines of consumer durables.[22]

As Salazar-Carrillo notes, a less aggregated analysis based on individual products undoubtedly would yield many more examples of advantages to be gained from specialization and trade.

In practice, of course, actual trade patterns under a Free Trade regime might change abruptly if exchange rate relationships were altered. In effect, therefore, Salazar-Carrillo's concept of absolute comparative advantage has economic and political components. Unfortunately, his effort to estimate purely economic comparative advantages is conducted at a much higher level of aggregation. Still, he suggests that Uruguay, Paraguay, Argentina, Brazil, and Chile all appear to have comparative advantages that should make them food exporters (not necessarily net exporters). Chile and Colombia have advantages in beverage trade. Colombia, Bolivia, and Ecuador have advantages in electricity, gas, and water. Argentina, Paraguay, and Venezuela could specialize in textiles. Argentina, Chile, and Venezuela have special advantages for the production of consumer nondurable goods. Venezuela, Argentina, and Colombia have advantages for consumer durables.[23]

For decades, many Latin American intellectuals have concluded that their nations must follow the path to advanced industrialization. For such products as transport equipment, heavy machinery and equipment, and construction goods, Salazar-Carrillo finds that Brazil and Mexico have the major advantages. Significantly, despite an already well-developed industrial sector, he finds that Argentina has advantages only for transport equipment.

These estimates accord well with more intuitive impressions. Argentina is blessed with its fertile pampas, as well as smaller districts suitable for specialized agriculture. But its stocks of ferrous minerals are inferior, its hydroelectric potential small and poorly located, its fossil fuels inadequate and expensive. Its economy would be most efficient if it were devoted to the production and processing of agricultural goods and to the fabrication of high-transport-cost consumer durables and nondurables. Such a strategy is quite dissimilar from the familiar "rich farmer" example of Denmark in that it stresses agriculturally based manufacturing. Brazil has both fertile agricultural lands and the potential to be the major supplier of industrial equipment to its neighbors in South America. Its primary regional competitor is Mexico, thousands of miles to the north. Smaller Chile has advantages in foods and beverages and in some nondurable consumer goods.

It is important to keep in mind that this assessment of comparative advantages does not preclude substantial national production of other products. The policy test of whether such production is appropriate is direct—does it require protection or subsidy? In contrast, protection and/or subsidies may be necessary to reach a competitive capacity in advantaged activities (the traditional "infant-industry" qualification to free trade). Understood in this context, Salazar-Carrillo's broad estimates of comparative advantage will be used as a standard against which to measure the strategies and results of historic political economy in Argentina, Brazil, and Chile.

Political Economy in Argentina, Brazil, and Chile

Having learned that all three members of Latin America's Southern Cone[24] are advantaged for food production, it is startling to find that Chile recently joined Brazil as a net food importer and that agriculture has not thrived in Argentina. Though a major exporter of wheat and meats, in some years Argentina governments have had to resort to domestic rationing to secure meat for export. Moreover, the output of meat and wheat per unit of land utilized is unimpressive. Until recent years, much of the twentieth century is best characterized as an era of agricultural decline.[25]

CHILE

Land ownership has been extremely concentrated in Chile throughout its history as a Spanish colony and independent country, with major reforms beginning in the 1960s.[26] Its major productive region requires irrigation during the growing season. Farther south, rains are more plentiful, many of the farms are smaller, and agriculture is more diversified. Grazing of beef gives way to dairy farms to supply the domestic market.

William C. Thiesenhusen provides the following summary of approximately five decades of agricultural expansion:

After 1875 and through the first third of the twentieth century, acreage used for agriculture doubled and irrigated acreage tripled in response to demands for agricultural products by expanding mining centers in northern Chile, by growing internal commercial centers, and by presently developed foreign countries which were not then self-sufficient in staples.[27]

During this period, cultivated acreage for grains expanded at the expense of acreage for grazing. Viniculture became important. But large estates combining extensive grain production and cattle ranching remained dominant. By the time the global depression hit Chile in the late 1920s, productivity gains had virtually ceased, and did not recover. By the mid-1950s, output per capita began to decline. The slowness and steadiness of agricultural decline are indicative of a fundamental change in incentives of some sort—in markets, taxes, investment opportunities, or general governmental climate —probably predating measurable changes by a decade or more.

Brian Loveman implies that that decline began much earlier, by the 1870s, as the wealthy elite began to lose interest in the productive enterprises that provided their wealth and instead turned to conspicuous consumption.

[s]umptuous houses and importation of luxurious European furnishing absorbed many of the windfall profits associated with wheat and flour exports, commerce, and mining. Santiago, Valparaiso, and even lesser central valley towns boasted new, lavishly decorated edifices. Upper-class gentlemen emulated the life styles of European capitalists and aristocrats. Prestigious social clubs and the National Agricultural Society (1869) brought together sociopolitical elites to decide matters of state and economy outside the public halls of the Congress. The best-endowed maintained haciendas near Santiago or Valparaiso as recreational retreats with ornamental gardens and well-furnished residences, or *casas de fundo*.[28]

Heavy spending also characterized the public sector, whose revenues quadrupled in the decade following the War of the Pacific.

Whatever the behavior of landlords and mining entrepreneurs, production and productivity increased well into the twentieth century, suggesting that one must look elsewhere for explanations. A more likely candidate is Chile's evolving political system and the policies it adopted to cope with underlying pressures from the poor and with a series of externally originated crises.

Although the potential for divisive conflict among agrarian, mining, and urban interests has been present since the 1880s, what evolved was a multiparty system of bargaining.[29] Formal coalitions governed during the Parliamentary Era (1891–1925); after that, a strengthened national executive was the key to coalition government involving shifting party and personal loyalties that persisted to the fall of Salvador Allende in 1973.[30] A second key was the steady expansion of the governmental bureaucracy. The Radical Party, initially a middle-class, individualist-oriented movement like its European namesake, promoted education guided by a government bureaucracy in control of finances. Its major contender during the early period was the National Party, which sought to promote international trade and to establish some control over it. Individual spokesmen and small parties pressed the claims of Chile's burgeoning poor, whose soaring numbers reflected the expansion of agriculture described above. These steps, it seems clear, were responses to particular constituencies or reactions to crises, not part of a conscious strategy.

One change, with major implications that seem favorable to agriculture, had occurred. Not only had fortunes been made in mining; the new sector came to support an expanding government:

[a]pparently the nitrate and copper booms and intermittent inflation cushioned the blow to Chilean agriculture. The burden of taxation was shunted to the mining sector and irrigation facilities for agriculture were financed out of the public coffers. This relieved landlords of the costs of bringing new land under the plow.[31]

Quite possibly, the decline in agriculture occurred even as these "relief" measures were implemented, more than offsetting them.

During the 1920s and 1930s, a public developmental strategy began to take shape. Its sources of inspiration ranged from the wartime (World War I) stopgap manufacturing to satisfy shortages, early nationalism inflamed by recurrent foreign exchange crises keyed to the boom-and-bust mining sector, domestic inflation, the deep depression that hit Chile with extraordinary severity, and the desire to provide jobs to better the lot of poor Chileans. Ideological in the sense of key ideas guiding government strategists, the program drew support from all sectors in its general aspects. The answer to Chilean ills, it was agreed, was industrialization, not expansion of traditional economic sectors. Government was to further this process by assisting private investors, channeling public investments, and guiding the direction of change.

The Constitution of 1925 greatly strengthened the role of the chief executive, especially over the national budget. The budget, in turn, provided the leverage for bureaucratic leadership. Radical political sociologist James Petras noted that:

the Chilean bureaucracy has been mainly responsible for providing services, such as roads, transport, schools, and so forth, that form a necessary "infrastructure" for socioeconomic development. More important, the bureaucracy's policies and regulations—tariffs, multiple-exchange controls, and so forth—and the way they are administered affect the social and political climate for development. This is important since policies aimed at can be and usually are rendered ineffectual by regulations that inhibit private or public entrepreneurial activity. The Chilean bureaucracy also plays an important entrepreneurial role, accounting for a high percentage of all investment.[32]

A shifting package of controls and incentives, coupled with direct investments under the control of the national bureaucracy, were the tools of a strategy that has come to be widely known as "import-substituting industrialization." Unfortunately, the strategy was pursued within a nation with a market so limited that it drastically reduced opportunities for efficiencies based on economies of scale. More seriously, perhaps, much investment was wasted on publicly supported projects directly counter to Chile's true potential.

Inauguration of the strategy of import-substituting industrialization under bureaucratic auspices coincides chronologically with the decline of Chilean agriculture. The essence of this strategy lies not in the commitment to industrialize but in the view that the aim of industrialization, apart from providing employment opportunities, should be to free the society from dependence upon imported manufactures and capital goods. Hence, instead of looking to current or prospective strengths, planners took guidance from areas of weakness. Inevitably, the result in most cases was inefficient, high-cost operations. Private investors were enticed to divert investments into these operations by the package of subsidies, protective trade measures, and privileged access to imports noted above.

Agricultural decline did not result from bureaucratic neglect or punitive policies, at least in any sense that would imply that power had been transferred from landed and mining elites to a new power bloc. Pierre R. Crosson documents an allocation of 30 to 40 percent of public credit per year to the agricultural sector during the 1950s.[33] Yet he concludes that a combination of unfavorable foreign exchange policies, coupled with credit supplies far below both demands and estimates of volume that could be employed profitably, counteracted efforts to help. By the 1950s, moreover, persistent inflation was diverting private finance into speculative hedges such as urban real estate and construction.

As demands upon the Chilean state grew during the 1950s, ways were sought to "milk" larger national shares from the foreign-dominated copper mines. Some small successes were quickly dissipated in new projects, in efforts to supply housing and other needs of the increasingly powerful poor, and by inflation. Not surprisingly, after three decades of agricultural decline, the landholding elite more and more became a target of reform as an unproductive drag upon the society. But as the drive to enact land reform gained recruits steadily, no consensus emerged regarding what should replace the large estates.[34]

The politics of the 1960s was the almost inevitable and tragic consequence of an industrialization strategy incapable of generating dynamism due to its incurable inefficiency and the antiquated agricultural sector left over from decades of neglect or worse. Nationalist rhetoric inflamed discussion of conditions under which copper would be available to foreign companies even as political parties competed frenetically for the hearts and votes of the rural and urban poor. Agricultural production lagged further, though it was partly offset by gains retained in the countryside by early beneficiaries of land reform. Rural violence became more common, with little hope that political institutions could contain these forces or structure participations by newly enfranchised Chileans.

By the time Salvador Allende had assumed the presidency as leader of a socialist coalition in 1970, decisive changes were necessary to reorient both industry and agriculture. But even the minority socialist coalition was torn by dissension on what was needed, agreeing only upon the desirability of the states or the workers taking control of enterprises and farms. Moreover, the need to fulfill campaign promises to transfer wealth immediately to the poor left little for investment. Whether Allende would have fallen when he did or shortly thereafter without United States' efforts to "destabilize" the regime, or whether he could have survived longer by negotiating with his competitors, are questions outside the scope of this essay. What is clear is that his program was devoid of guidelines on what Chile should produce and hence how a socialist Chile would differ in what it produced from what had gone before it.[35]

ARGENTINA

Argentina was settled from the west over the rugged Andes and was little more than a frontier outpost in the Spanish empire until the eighteenth century.[36] Reflecting that settlement pattern, its most traditional agricultural estates are in the interior, far from the rich soils of the Atlantic coastal plains. In the latter region, ranches and farms from an early point were oriented toward urban and European markets. By the mid-nineteenth century they were firmly linked to European—especially British—markets and sources of capital. Roads, railroads, and ports served this trade, and the introduction of beef refrigeration made it flourish. Ownership patterns were set, with most land in enormous estates. Family ownership frequently was interwoven with company ownership, in many cases including foreign partners.[37]

During the last half of the nineteenth century and well into the twentieth, Argentina's agrarian elite dominated the society. Native Indians were pushed off the inviting plains and virtually exterminated, making room for agriculture and especially for great cattle and sheep ranches. European labor was brought to the countryside to serve seasonal needs, and then frequently transported back to Europe. Urban packing plants, tanneries, and so forth were developed as natural extensions of an agrarian sector based firmly on comparative advantage. This agricultural specialization was an important requisite to Europe's rapid shift of resources into industrial activities. Because of this complementarity, it was relatively easy to finance infrastructure through the sale of government bonds abroad. Argentina's wealth grew rapidly, and its urban showpiece, Buenos Aires, became the "Paris" of Latin America.

In the first two decades of the twentieth century, the elite carefully managed extension of political participation to the middle class. It successfully blunted much of the challenge from an articulate socialist party centered primarily in Buenos Aires. But neither it nor Radical Party leadership could manage the dual challenge of new industrial groups created during World War I and the impact of the Great Depression. As the military tried to fill the gap, it was torn between the forces of fascism and nationalism on one hand and class polarization on the other.

During the crucial period from 1920 to 1950, the Argentine military contained a powerful faction that sided openly with the rural elite.[38] Assuming power in 1930 when the Radical presidency of Irigoyen collapsed, it sought to restore and stabilize the role of the trade-oriented rural elite, most notable by negotiating the controversial Roca-Runciman Treaty of 1933 with Great Britain. Seen from a global perspective of depression and trade wars, the treaty at least guaranteed markets. Domestically, it enflamed nationalism (because of prices accepted and loss of freedom of action) and widened the gap between elite and masses. In many other respects, military leadership was simply reactionary. This led to factionalism within the military.

In his important study of this era, Marvin Goldwert notes:

[s]ide by side with these traditional army values was a modern nationalism that called for autarchy, industrialization, and technical modernization. This nationalism, related to interest in a national war machine, included a drive to free the country from foreign influences and dependence on foreign markets. This strong technocratic drive also bolstered antagonisms in the German-trained army against the western democracies—England and the United States. Army nationalists suspected these nations of influencing Argentine politics through alliances with the oligarchy, and through large-scale investments in important sectors of the nation's economy. *They believed that the alliance between the oligarchy and the foreign interests was aimed at keeping the nation agricultural.*[39]

Since the collapse of the Radical Party left no domestic political group of any real significance to challenge military perspectives, the question became which military faction would rule Argentina.

Traditionalists governed the country from 1930 to 1943 amid growing corruption, further splits within political parties and even within the central labor confederation, and rapid migration of the rural poor to Buenos Aires. They were ousted by the opposing military faction in a coup that "was at once anti-oligarchical, antidemocratic, anticommunist, and sought no support from Argentine masses."[40] But one of its leaders, Juan D. Peron, did seek such support. He did not hesitate to use his posts in the military government (at one point he simultaneously was head of the Labor and War ministries and vice-president) to build a labor base loyal to himself.

When Peron attained dominance in 1946, he straddled a movement rooted in the nationalist-industrializer faction of the military and in the urban poor. He committed the country to a course of industrialization oriented away from traditional products and toward heavy industry, a strategic course that in key senses survived as the guide to government policies well into the 1960s.[41] The government became a vital force in credit allocation. It manipulated exchange rates and frequently resorted to exchange rate devaluation. It spent foreign currency reserves accumulated during the devaluation. It spent foreign currency reserves accumulated during World War II to purchase the aging railway system from the British, a decision that not only foreclosed alternative investment options but also forced the government as the employer into a negotiating position with a powerful trade union. Efforts were begun to help poor rural tenants through measures such as minimum contractural periods and rent controls. During the last years of control, wages and prices were frozen, allowing inflationary pressures to build.

The impact upon agriculture was sudden and drastic.[42] Crop production for the period 1950–54 was only 81.6 percent of its level during the depression years 1935–39. Apparently, in an effort to protect itself from governmental controls upon grain trade, the agricultural sector shifted toward cattle, and production of beef and mutton rose by 23.7 percent. Overall, per capita production declined as the agricultural sector stagnated. Depression levels of crop production (well below levels reached when Argentine agriculture led economic growth) were not equaled until 1960, by which time production had grown drastically. During the turbulent 1960s, with erratic but somewhat more favorable policies, gains averaged only 2 percent annually.

In recent years, following another Peronista interlude, the military has reversed some of its earlier policies, turning in some respects to market signals and substantially freeing the battered agricultural sector. Gains have been impressive, but it is far too early to judge whether they can be maintained and the traditional dynamism of the Argentine economy restored. Certainly it will be difficult to reduce commitments to well-developed heavy industry, despite Argentine dependence upon high-cost energy and mineral resources. The key may be avoidance of major new commitments, except where market signals indicate decent odds of success, rather than wasteful elimination of existing factories and consequent generation of political opposition.

One should not expect substantial or early reduction of governmental repression. Quite apart from periodic battles with guerrilla bands, the roots of commitment to the past strategy of industrialization are deep. It was, as noted above, a military faction that developed that strategy. Peron politicized it by appealing to and manipulating the urban masses in a classic example of "populist" leadership changing the course of the country in the absence of viable civilian party institutions. But Peron did not initiate the commitment away from an economy based upon comparative advantage, nor did his military and civilian successors abandon his basic approach to political economy. Hence when the Perons returned to govern the country briefly under a shaky truce with the military, few basic changes were apparent.

The nationalist-industrializer faction of the military remains an important alternative to the present effort to return to the guidance of market signals and comparative advantage. Any return to civilian rule probably would result in a similar reversion, since labor and industrial groups have been committed to that set of policies as well. Though not a real alternative, remnants of the religious-traditionalist military faction that sided with reactionist elements of the landowning sector in the 1930s not only remain but have added anti-communism applied indiscriminately to guerrilla, liberal, and intellectual groups to their appeal.

Politically, the odds for success of the present regime in imposing a change of course cannot be rated as high. Ironically, present military leaders may come to depend on support from the agricultural sector the same sector that preceding military factions systematically sought to weaken. For analysts generally, there can be little doubt that Argentina's four lost decades of agricultural stagnation, produced by government policies, is a crucial case study rich in lessons for viable food policies.

BRAZIL

Brazilian agricultural development during the Portuguese colonial era was a cross between the self-sufficient and inward-looking *latifundia* and commercial agriculture. It was dominated by tropical and subtropical plantations. The Steins show that by the seventeenth century the plantation owner dominated economic, political, and social life—subject only to limitations imposed by the imperial regime.[43] Plantation society epitomizes the dual society; the Brazilian variant has been analyzed extensively and provides the subject matter for such famous Brazilian novelists as Gilberto Freyre. In time, a peasantry intimately linked to the plantation economy emerged as a growing mass of rural poverty in northeastern Brazil.[44] Without question, plantation Brazil was an archetypal example of dependent development, though of mercantile rather than capitalist empire.

Development of Brazil's impressive agricultural resources in the central and southern parts of the country began much later and still contain major "frontier" areas. This second region comprises the dynamic core of the agrarian sector, the source of world leadership in coffee production and an area of growing importance for meats and grains. Located adjacent to Brazil's booming industrial centers, the coffee growers, ranchers, and agribusinessmen are very much a part of the elite that supports military rule today. It would constitute the alternative were the country to return to civilian government, providing only that it could work with the new industrial leaders. The emergence of this dynamic center in a political sense dates from 1930.

First under the monarchy and then under the so-called Old Republic, Brazilian government was dominated by rural patrimonial leaders and their urban commercial allies.[45] It by no means was stagnant. Mining, primarily for export, became important. The growing of coffee for foreign markets flourished. Deep involvement in international trade gave the Brazilian ruling elite a distinctly cosmopolitan orientation that easily transcended farm and plantation. Ranching filled some of the rich plains of southern Brazil. During the Old Republic, executive office rotated among leading states. Much more importantly, the state-based political machines insured that central government remained weak. Despite friction between the more dynamic coffee growers, ranchers, and new urban groups and the stagnant plantation sector, this arrangement persisted until a battle over the presidential succession.

In 1930, a disgruntled and ambitious rancher's son from the southern state of Rio Grande du Sul, Getulio Vargas, seized the presidency.[46] From 1930 to 1945, through the depression years and World War II, he managed to lead a shifting coalition of military and civilian groups, including coffee growers, farmers, and ranchers. After crushing the Marxist-led "Prestes revolt" he proceeded to suppress ruthlessly the small labor organizations that had been developing in the cities, eliminating any important political pressure from urban workers.

Though it is not clear that Vargas had a particular economic design in mind, he was conscious of political realities. As he began to build the power of central government in Brazil, he used that power to promote industrial development without doing so at the expense of agriculture.[47] Most notable, coffee markets were stabilized through government purchases and related protective measures. Vargas's strategy benefited greatly from the decision to collaborate as a close ally with the United States during World War II. This brought airfields and port developments, and helped create an atmosphere conducive to foreign investment. Though foreign investors took advantage of the local market protection afforded by wartime shortages, their investment decisions were not keyed to promises of government subsidy and protection. The most important decision made by the Vargas regime was to create a basic iron and steel industry. The natural resources were available. Brazilian governmental involvement made it possible to link domestic investors together with sources of international financial support, thereby creating the Volta Redonda complex in a comparatively short time and laying the basis for heavy industry today.

Vargas's strategy, as noted, was to industrialize without harming agriculture. Access to foreign financing, private and public, provided the necessary resources to avoid diverting funds drained from the agricultural sector. Nevertheless, the agricultural boom associated with development of coffee had run its course. The northeastern plantations remained stagnant. Although some grazing lands in south and central Brazil were converted to products that yielded much higher returns, such as sugar, productivity elsewhere lagged. Most grains in output came from development of new, rich soils on the frontier, rather than improved practices. There is not doubt that resources from agricultural sector were diverted into the more lucrative possibilities available in the booming industrial sector.

By the time Vargas was forced to yield power in 1945, he had given the government bureaucracy key economic roles in the steel industry and in the lagging petroleum industry (one of the weak spots in Brazil's bounteous resource inheritance). Vargas had attained some personal popularity among urban laborers, and had begun the process of creating a government-labor linkage under the Labor ministry unmediated by powerful union organizations. He had succeeded in attracting technical skills from the private sector and from the military (engineering, etc.) into government. And he had begun a largely repressionless (with exceptions noted above) shift toward industrialization without alienating traditional elites. The key to this transformation, and to increasing benefits for laborers, was the attraction of foreign investment and the willingness of Brazilians to use their wealth for productive investment rather than to provide a lifestyle of indolent wealth.[48]

Following a brief period in which a portion of the military connived with the old party leadership, Vargas returned to the presidency in 1950. His role became increasingly populist as he sought the mantle of labor leadership and adopted some of the nationalist rhetoric of urban spokesmen.[49] Like his successors, Kubitschek, Quadros, and Goulart, he expanded the state role by sponsoring expensive public works, manipulating tax incentives and exchange rates, licensing favored capital imports, and seeking foreign financing for rapid economic growth. The basic style of "clientelist" politics, that left enormous discretion to the executive and bureaucracy, was by now established.[50] But more than his successors, Vargas seemed to understand the need for foreign investment and cooperation with the Brazilian private sector. That meant a general commitment to comparative advantage and efficiency, not Brazilian self-sufficiency and autonomy or autarky. Thus, for example, iron ore exports expanded as the domestic industry was being created. Resource development was financed largely by private foreign investors in response to market needs. This conserved foreign public loans, Brazilian government funds, and domestic private resources for the industry and subsequent manufacturing utilizing steel products.

Vargas's successors were less willing to make distinctions and much less willing to compromise with domestic and foreign financiers. They fanned Brazilian nationalism, posturing that at a very early stage of development Brazil already offered the developing countries a "third way" between Soviet and American models. They sought to fill the gap left by uncertain foreign investors with funds squeezed from coffee exports, thereby risking opposition from coffee growers and making them less willing to cooperate in joint ventures.[51] In later years, with executive support or at least tolerance, government bureaucrats encouraged agrarian unrest in the northeast, putting it into direct conflict with the old plantation elite and its supporters in the military. Crucially, the conditions for a steady flow of foreign financing, private and public, were not met. The result was steadily worsening inflation. Because that inflation was relatively mild (except just before the fall of Goulart) and brief, one result was a rise in food production.

The political economy of the harsh military regime that seized power in 1964, crushing agrarian and urban unrest and destroying most existing political organizations, is a return to the basic formula of Vargas. An important addition to the strategy is the encouragement of foreign investment in agriculture, especially in developing modern beef operations for export to European markets and in converting acreage to such crucial products in the world food diet as soybeans. The formula of blending state capitalism with domestic and foreign participation continues. International bargaining over investments, capital-supply contracts, and credits has become increasingly sophisticated, reflecting both experience and recognition of Brazilian assets. The economic result thus far has been very rapid growth of industry and agriculture in what is without question one of the most dynamic economies in the world.[52]

Brazil's dynamic development is heavily dependent upon its foreign ties for financing. Its foreign debt has soared, in part because of soaring costs for petroleum. In an interesting effort to reduce petroleum dependency by shifting to gasohol and alcohol as auto fuels, the regime has stimulated long-dormant plantations. For the first time in this century, the traditional elite of the northeast has a stake in the regime that is not defensive. But it would be misleading to characterize Brazil as dependent in the mode of primary product supplier to more advanced economies. Like those of the

United States, its rich agricultural resources will yield exports regardless of industrialization. More and more of its mineral resources are used in domestic industry, both for domestic and for export markets. In short, though some distance from having completed the transition, Brazil is on the verge of entering the category of developed industrial economy.

The Brazilian model, especially its political economy, is being emulated by both Argentina and Chile. Its key features are summarized by Albert Fishlow:

> Along with the emphasis upon market signals and free market determination, the role of the state in the economy has been enhanced. Government has frequently come to invest more rather than less, usually in new sectors requiring large amounts of capital and in association with foreign investors. Even when they have not partici-pated in production directly, governments have become more powerful regulators of business activity, and of foreign investment.
>
> The convergence within the region to the new outward oriented and capitalist economic model in the past decade has been impressive.
>
> Traditional commodities, including petroleum, have declined from three-fifths of exports in the 1960s to less than 50 percent in 1977; manufactures alone have doubled from 10 to 20 percent of exports.[53]

Again, the aggregate figures miss some important trends. Many of Brazil's exports, especially nontraditional ones, go to its neighbors in Latin America. This is not the result of regional organizations. Instead, it simply reflects the advantages that geographical proximity add to reasonably efficient produc-tion of heavy products such as rolled steel. In time, Brazil should be the primary exporter of automobiles and similar products to its neighbors. Regional organizations could speed this process, but to do so will require much clearer recognition by Argentine leaders that each can benefit from specialization. Such understanding is much more difficult to achieve and gain public support for than such limited agreements as the ones for joint hydroelectrical projects achieved thus far.

Conclusion

This essay has traced the analytical and real-world advantages of approach-ing food policy from the perspective of national comparative advantage. Thus, it challenges scholars of regional organizations to go beyond charter provisions and descriptions of changes in trade and investment flows. It suggests that changes can be sustained only if they build upon respective advantages and encourage specialization defined by those advantages. Other-wise, they will be the victims of waning commitment because they have no foundation in mutual self-interest. This suggestion is both political and economic, since it implies the necessity of analyzing national development strategies. Finally, as each of the cases demonstrates, a national commitment to being guided by market signals and comparative advantage may also entail a major role for the state in the economy. Whether this phenomenon in Latin America is a necessity of "late" development, or a matter of national pref-erence is an intriguing, unanswered question.

Notes

1. For a representative selection of contemporary work on food policy, see the special issue of *International Organization*, vol. 32, no. 3 (Summer 1978), edited by Raymond Hopkins and Donald J. Puchala under the issue title of "The Political Economy of Food." Much of this literature is rich in both insights and data on frequently neglected topics, despite the flaw critiqued here.

2. For an excellent description from a critical perspective, see Brian Loveman, *Chile: The Legacy of Hispanic Capitalism* (New York: Oxford University Press, 1979), especially chapters 5 and 6.

3. See, e.g., Gilberto Freyre's classic *The Masters and the Slaves* (New York: Alfred A. Knopf, 1946). Shepard Forman's invaluable study *The Brazilian Peasantry* (New York: Columbia University Press, 1975) argues that, by the time the master/slave plantation civilization approached its zenith in the early nineteenth century, a peasantry integral to it was already well developed.

4. The most extensive application to the region is Charles Anderson's *Politics and Economic Change in Latin America* (New York: D. Van Nostrand, 1967). The ten countries examined by Anderson do not include the three surveyed in this essay.

5. Gary Wynia, *The Politics of Latin American Development* (New York: Cambridge University Press, 1978).

6. Ernest Feder, *The Rape of the Peasantry* (Garden City, N.Y. Anchor Books), p. 291.

7. Advocates of "appropriate technology" correctly insist that when labor is cheap it makes little sense to displace it with expensive capital-intensive machinery. That does not reduce the necessity of utilizing capital to improve productivity, nor can it be permitted to obscure the reality of massive unemployment and underemployment in low-productivity agrarian societies. Still classic is Ragnar Nurske, *Problems of Capital Formation in Underdeveloped Countries* (Oxford: Basil Blackwell, 1958).

8. Robert Tucker, *The Inequality of Nations* (New York: Basic Books, 1977).

9. "Equity in the South in the Context of North-South Relations," in Albert Fishlow, Carlos F. Diaz-Alejandro, Richard R. Fagen, and Roger D. Hansen, *Rich and Poor Nations in the World Economy* (New York: McGraw-Hill, 1978), pp. 163–214.

10. Norman Nicholson and John Esseks, "The Politics of Food Scarcities in Developing Countries," *International Organization* 32, no. 3 (Summer 1978): 692.

11. In the abstract, many other potential options exist, such as the use of some resources for subsidies and others for investments, especially if adequate foreign funds are made available. For a variety of reasons, such balanced strategies rarely have been sustained in practice.

12. Garrett Hardin, "The Tragedy of the Commons," *Science* 162 (1968):1243–48.

13. The contributors to Garrett Hardin and John Baden, eds., *Managing the Commons* (San Francisco: W. H. Freeman, 1977), developed the underlying issues for a variety of resources, circumstances, and legal settings.

14. The World Bank, *World Development Report, 1978* (New York: Oxford University Press, 1979), table 1, pp. 76–77.

15. Hopkins and Puchala, "The Political Economy of Food," pp. 586–87.

16. James E. Austin, "Institutional dimensions of the malnutrition problem," *International Organization* 32, no. 3 (Summer 1978):812–13.

17. The World Bank, *World Development Report*, p. 52.

18. Ibid., p. 77.

19. Joseph Grunwald and Philip Musgrove, *Natural Resources in Latin American Development* (Baltimore, Md.: Johns Hopkins Press, 1970).

20. Dennis Pirages, *Global Ecopolitics* (North Scituate, Mass.: Duxbury Press, 1978), seeks to reduce slipperiness by introducing natural science concepts of natural and current carrying capacity. The solution is only in abstraction. Natural carrying capacity in principle assumes optimum techniques and hence is subject to constant adjustment to reflect discoveries, inventions, and innovations. Current carrying capacity appears to assume isolated national systems, since the necessity of international exchanges is equated with vulnerability. This

definition therefore slips in a bias toward national self-sufficiency under the mantle of scientific necessity.

21. Jorge Salazar-Carillo, *Prices and Purchasing Power Parities in Latin America 1960–1972* (Washington, D.C.: Organization of American States for the Eciel Program, 1978).

22. Ibid., p. 9–13.

23. Ibid., pp. 9–20.

24. Strictly speaking, Uruguay and perhaps Paraguay should be included if the intent is geographical.

25. See, especially, Darrell F. Fienup, Russell H. Brannon, and Frank A. Fender, *The Agricultural Development of Argentina* (New York: Praeger Publishers, 1969).

26. This summary of Chilean agricultural evolution draws upon the interpretations of William C. Thiesenhusen, *Chile's Experiments in Agrarian Reform* (Madison: University of Wisconsin Press, 1966); Loveman, Markos Mamalakis and Clark W. Reynolds, *Essays on the Chilean Economy* (Homewood Ill.; Richard D. Irwin, 1965); Grunwald and Musgrove, *Natural Resources;* and Robert D. Kaufman, *The Politics of Land Reform in Chile, 1950–1970* (Cambridge, Mass.: Harvard University Press, 1972). The author's interpretation differs substantially from those of authors cited.

27. Thiesenhusen, *Chile's Experiments*, p. 13.

28. Loveman, *Chile*, p. 165.

29. Federico Gil, *The Political System of Chile* (Boston: Houghton Mifflin, 1966), includes a clear description of the evolution of coalition politics and of constitutional changes.

30. Peter A. Goldberg, "The Politics of the Allende Overthrow in Chile," *Political Science Quarterly* 90, no. 1 (Spring 1975):93–116, includes useful discussion of coalitional politics in the 1960s and 1970s.

31. Thiesenhusen, *Chile's Experiments*, p. 14.

32. James Petras, *Politics and Social Forces in Chilean Development* (Berkeley, Calif.: University of California, 1969), p. 297.

33. Pierre R. Crosson, *Agricultural Development and Productivity: Lessons from the Chilean Experience* (Baltimore, Md.: Johns Hopkins Press, 1970), pp. 12–13. Crosson in subsequent chapters carefully contrasts apparent opportunities for profit with the flow of actual investments.

34. Kaufman provides a detailed discussion of conflicting land reform programs and of the politics underlying them.

35. Especially valuable interpretations of underlying trends that also discuss widely varying treatments of the period are Paul E. Sigmund, *The Overthrow of Allende and the Politics of Chile, 1964–1976* (Pittsburgh, Pa.: University of Pittsburgh, 1977); and Arturo Valenzuela, *Chile: The Breakdown of Democratic Regimes* (Baltimore, Md.: Johns Hopkins Press, 1978).

36. An insightful interpretive study of the economic, social, and political history in the tradition of the nineteenth-century writer, educator, and president Domingo F. Sarmiento is James R. Scobie, *Argentina: A City and A Nation* (New York: Oxford University Press, 1964). More specialized is Peter Smith, *The Politics of Beef in Argentina* (New York: Columbia University Press, 1969).

37. See especially Alberto Ciria, *Parties and Power in Modern Argentina*, trans. Carlos A. Astiz and Mary McCarthy, (New York: New York University Press, 1974), pp. 247–53.

38. The leading study is Marvin Goldwert, *Democracy, Militarism, and Nationalism in Argentina, 1930–1966* (Austin, Tex.: University of Texas Press, 1969).

39. Ibid., p. 64.

40. Ibid., p. 81. Additional helpful background on the availability of the urban masses to any serious organizational effort that demonstrates the absence of any such effort by the Radical Party during the 1916–1930 period of ascendancy is provided in David Rock, *Politics in Argentina, 1890–1930* (New York: Cambridge University Press, 1975).

41. On Argentine political economy since 1943, see Aldo Ferrer, *The Argentine Economy* (Berkeley, Calif.: University of California Press, 1967); Jean-Claude Garcia-Zamor, *Public Administration and Social Changes in Argentina, 1943–1955* (Author's copyright, 1968); Darrell F. Fienup, Russell H. Brannon, and Frank A. Fender, *The Agricultural Development of Argentina* (New York: Praeger Publishers, 1969); Carlos F. Diaz-Alegandro, *Essays on the History of the Argentine Republic* (New Haven, Conn.: Yale University Press, 1970); Richard D. Mallon with

Juan V. Sourrouille, *Economic Policymaking in a Conflct Society: the Argentine Case* (Cambridge, Mass.; Harvard University Press, 1975); and Laura Randall, *An Economic History of Argentina in the Twentieth Century* (New York: Columbia University Press, 1978).

42. Figures cited are from Fienup, Brannon, and Fender, *The Agricultural Development of Argentina*, p. 62.

43. Sidney J. and Barbara H. Stein, *The Colonial Heritage of Latin America* (London: Oxford University Press, 1970), p. 66.

44. Forman, especially the provocatively titled second chapter, "Beyond the Masters and the Slaves."

45. The text by political scientist Riordan Roett provides a brief historical description built around the analytical construct of patrimonial rule in *Brazil: Politics in a Patrimonial Society* (Boston, Mass.: Allyn and Bacon, second edition, 1979).

46. Thomas E. Skidmore includes substantial coverage of the changing political economy in *Politics in Brazil, 1930–1964* (London: Oxford University Press, 1967); Grunwald and Musgrove, *Natural Resources*, pp. 187–212, provide valuable data and commentary.

47. Agricultural policy and development is covered in Janet D. Henshall and R. P. Momsen, Jr., *A Geography of Brazilian Development* (London: G. Bell and Sons, 1976).

48. Two important studies largely devoted to later years give valuable analyses of the Brazilian situation in the mid-1940's. See Nathaniel H. Leff, *Economic Policy-Making and Development in Brazil, 1947–1964* (New York: John Wiley, 1968); and Alfred Stepan, *The Military in Politics: Changing Patterns in Brazil* (Princeton, N.J.: Princeton University Press, 1971).

49. The literature on populism as a political and as a socioeconomic phenomenon is voluminous. Its primary value appears to be critical, demonstrating how limited democracy in Latin America has been in practice and hence revealing the difficult role faced by leaders such as Peron and Vargas as newly politicized urban dwellers began unstructured political participation. The major change in Vargas's behavior from suppressor of labor leadership to advocate for urban workers reflected the necessity of balancing a more complicated coalition in the 1950s. The military organization had developed enormously during World War II, during which it sent troops to Italy and worked closely with the United States. The number of urban workers also had multiplied. Had Vargas retained his position of 1930, he would be regarded as reactionary. See the interesting treatment of Peron and Vargas as populists (with the designation applied only to the latter portion of the Vargas era) in Wynia, chapter 6. For representative collections that show both the insights and tendency toward arid jargonizing in literature, see Frederick B. Pike and Thomas Stritch, eds., *The New Corporatism* (Notre Dame, Ind.: University of Notre Dame, 1974) and James M. Malloy, ed., *Authoritarianism and Corporatism in Latin America* (Pittsburgh, Pa.: University of Pittsburgh, 1977). The latter collection contains examples from the "dependency" perspective.

50. Leff, *Economic Policy-Making and Development* pp. 118–31.

51. Ibid., p. 189.

52. Excellent but becoming dated by rapid change is Donald E. Syvrud, *Foundations of Brazilian Economic Growth* (Stanford, Calif.: Hoover Institution Press and American Enterprise Institute for Public Policy Research, 1974).

53. Subcommittee on Western Hemisphere Affairs, Committee on Foreign Relations, United States Senate, *Hearings*, 3, 4, and 6 October 1978, p. 60.

6

East and Southeast Asian Food Systems: Structural Constraints, Political Arenas, and Appropriate Food Strategies

DAVID N. BALAAM

Food systems in East and Southeast Asia highlight a variety of political and economic factors that influence the occurrence of hunger and the effectiveness of food politics. Because hunger is directly related to poverty, this region's food problems mirror those of the entire world based on levels of economic development achieved by nations in the region.[1] Most of the poorer nations in the region are committed to generating economic growth, which necessarily incorporates a consideration of hunger problems and food policies in their development strategies.

However, conditions of overpopulation, underproduction, hunger, malnutrition, and even intentional starvation are common and continue to defy solutions. As much as 80% of Asia's population, most of it in rural areas, is malnourished.[2] Because of these conditions and in order to stimulate economic growth, all Asian nations regulate their economies and markets in ways they believe are conducive to development, political independence, or security. This has important implications for hunger relief and the effectiveness of development strategies.

Given four similar kinds of constraints faced by the region's poorer nations—(1) natural limitations, (2) pressures to industrialize and/or increase

This research was partially funded by a University of Puget Sound Faculty Enrichment Grant. Many thanks for critical comments of an earlier draft are due to Michael Carey, Richard Moore, and Lynn Scarlett. Thanks for manuscript preparation are due to Nancy Rees, Dan Pearson, and Florence Phillippi. None of these people are responsible for my mistakes.

agricultural output, (3) external political and economic linkages to the international system, and (4) security concerns—the problem is whether they can adopt an appropriate food and development strategy that relieves hunger. This chapter argues that these constraints narrow the range of policy options available to policymakers, forcing poorer nations to adopt a traditional outward-looking strategy that emphasizes growth but that also skews income and perpetuates hunger.

Scope

Asian food systems are similar enough to warrant analysis by the regional approach. Historically, people in the region have attempted to separate themselves from the rest of the world with the result that outsiders' interest in the region varied from curiosity about old cultures behind long walls to attempts to open trade doors to "oriental" markets. However, changes in the distribution of wealth and other sources of power in the international system brought on by the industrial revolution significantly changed Asia's relationship to the rest of the modern world and Asia's image of itself.[3]

Some countries, such as Japan, took advantage of the situation, realizing that no choice was left but to learn from and accept Western influence in order to develop and secure itself from the West. Thus, Japan continues to defy characterization as a Western industrialized nation despite acquired Western eating habits and materialistic values. China and poorer nations at the other end of the development continuum have not fared so well. Their physical traits, cultural values, and relationship to the developed nations deterred development and hunger relief, except in small enclaves that maintained and promoted neocolonial and imperial linkages to the advanced regions.

This chapter focuses on three subregional groups of East and Southeast Asian nations: high- to middle-income market economies, a number of centrally planned economies, and five low-income market economies.[4] Particular attention is paid to food production (see Table 6.1), distribution, and consumption (see Table 6.2) trends within these nations. The prospects for development of a regional organization are also discussed.

Since solutions to Asian food problems are more complicated than simply increasing rice production, this work also focuses on food-pricing, market, and trade policies formulated in the political arenas of these three subregional groups of nations. The boundaries of national political arenas are fixed by underlying structural conditions (perceived by policymakers as givens) or in accordance with the success or failure of past policies.[5] Structural conditions include, among other things, food production and population growth rates, industrial output levels, land tenure systems, cultural norms, and the like. They limit the range of options available to policymakers and must be reconciled with features of the political arena—the decision and policy-making structure and process (which is most often authoritarian), different ideological and economic outlooks, regime support, and national wealth and welfare and security goals.

Table 6.1 Food Production in East and Southeast Asia

| | Average annual growth rates | | | | Indices of food production (1969-71 = 100) | | | | | |
| | Total food production | | Per capita food production | | Total | | | Per capita | | |
	1961-70	1970-76	1961-70	1970-76	1973	1975	1977	1973	1975	1977
High and middle-income market economies										
Japan	3.2	2.3	2.1	0.9	101	110	108	97	103	99
Hong Kong	-5.5	-5.1	-7.5	-6.5	–	–	–	–	–	–
South Korea	4.5	2.6	1.9	0.6	–	–	–	–	–	–
Taiwan	–	–	–	–	–	–	–	–	–	–
Singapore	4.3	5.5	2.1	3.9	–	–	–	–	–	–
Centrally planned economies										
PRC (China, Peoples Republic of)	–	–	–	–	110	118	123	105	109	110
Vietnam	–	–	–	–	107	112	116	101	101	100
Cambodia	3.7	-10.6	0.9	-12.9	61	67	71	57	59	59
North Korea	–	–	–	–	117	135	156	108	118	130
Laos	6.8	1.6	4.4	0.6	105	112	113	98	100	97
Low-income market economies										
Philippines	3.2	5.5	0.0	2.1	114	128	140	103	108	111
Indonesia	3.1	3.6	0.6	0.9	117	120	126	109	106	105
Malaysia[a]	5.2	6.9	2.3	4.0	118	127	134	109	110	111
Burma	1.3	2.4	-0.9	-0.1	103	106	112	96	94	95
Thailand	3.8	4.3	0.7	1.0	123	128	124	111	109	99

[a] These figures are for Malaysia Peninsula.

Source: U.N. Statistical Yearbook, 1978, and FAO, The State of Food and Agriculture, 1978 (Rome).

108

Table 6.2 Per Capita Dietary Energy Supplies in Relation to Nutritional Requirements, Selected Developing Countries and Areas

	Requirements		As percentage of requirement	
	Kilocalories per capita per day	1974 Supply	Average 1969-71	1974
High- and middle-income market economies				
Japan	—	2,835	—	121
Hong Kong	2,290	2,533	114	111
Singapore	2,300	2,819	118	87
South Korea	2,350	2,630	115	115
Taiwan	—	2,780	—	119
Centrally planned economies				
PRC	—	2,330	—	99
Vietnam	2,160	2,397	105	107
Cambodia	2,220	1,894	100	85
North Korea	—	2,691	—	113
Laos	2,220	2,090	95	94
Low-income market economies				
Philippines	2,260	1,971	86	87
Indonesia	2,160	2,126	91	99
Burma	2,160	2,223	101	103
Thailand	2,220	2,382	103	107
Malaysia	2,240	2,574	111	115

Source: FAO, *The State of Food and Agriculture 1978* (Rome).

The region contains two popular and distinct models of development that have not proven entirely adaptable to poorer nations in the region. Japan adopted a traditional "Industry First" outward-looking strategy that balanced economic growth and the intersectoral exchange process[6] with natural constraints and political objectives. On the other hand, China's popular socialist experiment until recently emphasized an "Agriculture First" inward-looking self-sufficiency approach to development whereby leaders have consciously tried not to foresake the ideological goal of social justice for economic growth.

Whether these two development models are appropriate for other East and Southeast Asian or even non-Asian developing nations remains an issue. Analysts from a variety of backgrounds have begun to doubt their appropriateness to local conditions and contribution to either food production and distribution or economic development.[7] Despite efforts to find alternatives to a model thrust upon them by their colonial experience, the poorer nations of the region have relied almost exclusively on the traditional (Japanese) development model, which results in skewed incomes and perpetuates hunger. The centrally planned economies have relied on the socialist distribution development strategy, but they have not been able to stimulate economic growth. Many national leaders continue to rely on development strategies that emphasize industrial production, even if to create new input technologies for the agricultural sector. The most common result, however, is still an asymmetrical distribution of income and food in favor of a wealthier minority while the poor (society's greatest numbers) never actually improve their dietary conditions.[8]

High- and Middle-Income Market Economies

The rural-urban intersectoral exchange process in Japan, Hong Kong, Singapore, Taiwan, and South Korea is either near complete or well on its way. All the high-income countries except South Korea have per capita incomes of at least $1,000 and an increase in population growth rates of 2.0% per year or less (see Table 6.3). An increasing percentage of population in these countries is engaged in urban and industrial sector employment activities (see Table 6.4).

These countries have some of the fastest-growing economies in the world. They all focus on increasing industrial and manufactured exports. During the 1960s, rates of manufacturing and industrial growth in Taiwan and Hong Kong were related to the increased value of textiles, in Japan to metal products, and in South Korea to transport equipment.[9] These countries employ a variety of import substitution and import tariff measures to stimulate local production and protect infant industries.

Consumer food costs in these countries have been kept low to encourage investment and market activity in the urban-industrial sector.[10] At the same time, the proportion of agricultural products as part of gross domestic product (see Table 6.5), export goods, and total investment in agriculture has declined. Because of their limited growing areas, high rates of population per

Table 6.3   Population and Income Growth

	Population mid-1977 (millions)	Population growth rates (%) 1960-70	1970-77	GNP per capita Amount 1977 (US $)	Real growth (%) 1970-77
High- and middle-income market economies					
Japan	113.2	1.0	1.2	6,510	3.6
Hong Kong	4.5	2.5	2.0	2,620	5.8
Singapore	2.3	2.4	1.5	2,820	6.6
South Korea	36.0	2.4	2.0	980	7.6
Taiwan	16.8	2.7	2.0	1,180	5.5
Centrally planned economies					
PRC	885.6	1.9	1.3	410	4.5
Vietnam	50.6	3.1	3.1	–	–
Cambodia	8.4	2.8	2.5	–	–
North Korea	16.7	2.8	2.6	680	5.3
Laos	3.2	2.2	1.1	90	–
Low-income market economies					
Philippines	44.5	3.0	2.7	460	3.7
Indonesia	133.5	2.2	1.8	320	5.7
Burma	31.5	2.2	2.2	140	1.3
Thailand	43.8	3.1	2.9	430	4.1
Malaysia	13.0	2.9	2.7	970	6.5

Source: International Bank for Reconstruction and Development, *World Bank Atlas,* 1979.

square mile, and comparative advantage in trade, these nations are not likely to become, nor do they all desire to become, self-sufficient food producers. Instead, they depend on food imports to meet income demands. People in the high- and middle-income market economies of this region generally eat better-quality and more expensive food than people in the poorer nations and spend a smaller share of their income for food. A recent U.S. study concludes that these countries have absolutely increased their calorie con-

Table 6.4   Structure of Population

	Agriculture 1960	Agriculture 1977	Industry 1960	Industry 1977	Service 1960	Service 1977	Urban population as % of total population 1960	Urban population as % of total population 1975
High- and middle-income market economies								
Japan	33	14	30	37	37	49	62	75
Hong Kong	8	2	52	57	40	41	89	90
Singapore	8	2	23	32	69	66	100	100
South Korea	66	45	9	33	25	22	28	49
Taiwan	56	34	11	27	33	39	36	51
Centrally planned economies								
PRC	75	63	15	24	10	13	19	23
Vietnam	81	70	5	9	14	21	15	20
Cambodia	82	75	4	4	14	21	11	13
North Korea	62	51	23	32	15	17	40	55
Laos	83	80	4	6	13	12	8	11
Low-income market economies								
Philippines	61	51	15	15	24	34	30	34
Indonesia	75	60	8	12	17	28	15	18
Burma	68	55	11	19	21	26	19	25
Thailand	84	77	4	8	12	15	12	14
Malaysia	63	44	12	20	25	36	25	30

The column headers group as: *Percentage of labor force in* (Agriculture, Industry, Service) and *Urban population as % of total population*.

Source: International Bank for Reconstruction and Development, *World Development Report, 1979,* August 1979.

Table 6.5   Distribution of Gross Domestic Product (percentage)

	Agriculture 1960	Agriculture 1977	Industry 1960	Industry 1977	Manufacturing[a] 1960	Manufacturing[a] 1977	Services 1960	Services 1977
High- and middle-income market economies								
Japan	13	5	42	41	33	30	45	54
Hong Kong	4	2	34	31	25	26	62	67
Singapore	4	2	18	35	12	25	78	63
South Korea	40	27	19	35	12	25	41	38
Taiwan	28	12	29	46	22	37	43	42
Centrally planned economies								
PRC	—	—	—	—	—	—	—	—
Vietnam	—	—	—	—	—	—	—	—
Cambodia	—	—	—	—	—	—	—	—
North Korea	—	—	—	—	—	—	—	—
Laos	—	63	—	13	—	3	—	24
Low-income market economies								
Philippines	26	29	28	35	20	25	46	36
Indonesia	54	31	14	34	8	9	32	35
Burma	33	47	12	11	8	9	55	42
Thailand	41	27	18	29	11	20	41	44
Malaysia	37	26	19	29	8	18	45	45

[a] Manufacturing is shown separately (even though it is part of the industrial sector) because it is the most dynamic part of the industrial sector.
Source: International Bank for Reconstruction and Development, *World Development Report, 1979,* August 1979.

sumption and that South Korea generally represents nations where "average nutritional levels comfortably exceed minimum health requirements."[11]

JAPAN

In many respects Japan is a good example of a nation that has solved some of its food production problems. Japan is nearly self-sufficient in a number of commodities including rice, eggs, meats, and fruits. Imports of wheat, barley, rye, and beef have increased recently in response to rapidly rising per capita incomes and changes in eating habits due to Western influence. However, two food problems of political and economic importance also faced by other advanced industrialized nations remain. First is the rising cost of food reflected in high consumer prices (see Table 6.6) and the cost of government price supports that make for rice trade prices three times the international market price. These problems directly conflict with a second problem. Japan is the world's third richest and fourth most active trading nation. A new concern has emerged about Japan's growing vulnerability to other nations. Increasing interdependence, especially with other advanced nations, has further politicized Japan's food and agriculture policies and made their management in the Diet and in such international forums as GATT difficult to reconcile.

STRUCTURAL CONDITIONS: A HISTORICAL PERSPECTIVE

Structural conditions played a large part in influencing Japan's development and agriculture policies over the last one hundred years. The Meiji restoration of 1868 marked a turning point when leaders chose and later successfully

Table 6.6   Annual Changes in Consumer Food Prices (percentage)

	1960 to 1965	1965 to 1970	1970 to 1978
Japan	7.2	6.1	13.0
South Korea	18.3	12.5	16.8
Cambodia	2.7	6.7	112.8
Laos	39.0	4.0	40.9
Burma	—	2.9[a]	21.0
Indonesia	—	100.0	25.2
Malaysia, Peninsular	0.6	0.4[a]	10.4
Philippines	6.8	5.2[a]	20.1
Thailand	2.0	4.2	11.9

[a] 1967-70

Source: FAO, The State of Food and Agriculture, 1978, Rome, Italy.

implemented "an efficient strategy for agricultural development."[12] Through the selective borrowing of (Meiji) technology, including fertilizer-responsive rice varieties, improved cultivation methods, and land improvement activities that complemented Japan's system of small-scale farming, production increased and efficiency improved.

Until the 1930s Japanese agriculture financed the growth of the industrial sector. Because few demands were made on the growing urban-industrial sector, scarce capital and foreign exchange accumulated and stimulated more savings and investment in industrial, manufacturing, commercial, and transport activities. Between 1888 and 1892 the government acquired 85 percent of its revenue from agriculture while "the reverse flow of government funds to agriculture was extremely limited."[13]

Development did not occur smoothly in Japan, however. After World War I the international competitive position of the yen remained overvalued, and growing world inflationary conditions forced the government to adopt deflationary policies, causing a slowdown in industrial output and an increase in urban unemployment. Cities could not absorb the people who left the agriculture sector and food production did not keep up with demand. Therefore, Japan had to import rice from Korea and Taiwan. This, in turn, cut into local production markets, limited domestic output, and led to higher rice prices. Rice riots were touched off in 1918 because consumers were unwilling to buy more expensive (i.e., Western) food items and held on to rice as their staple food.

Later, as Japan became more integrated with international markets, American and European trade barriers against Japanese products further contributed to this situation. When food production recovered and began to outrun urban demand it became necessary to replace unavailable domestic markets with external outlets. Imperialistic foreign economic policies were easily rationalized. Japan's intersectoral exchange patterns were linked to and sustained by its influence in Asia, which partially explains Japan's attempt to extend influence there today (discussed below). But before and during World War II the imbalance between industrial growth and agricultural output effectively contributed to "the successful seizure of power by the militarist clique and ended with the nation's defeat in World War II."[14]

After the war Allied Occupation Headquarters initiated land reform measures to stimualte production and provide grass-roots support for Japan's new democracy.[15] But a gradual accumulation of rice surpluses contributed to a decline in the absolute numbers and income of farmers at a rather drastic rate. Since the mid-1950s the government has sought to control production through a number of measures, because Japanese farmers have continually shown a capacity to produce more rice than they have a market for. Although a number of these measures have contributed to a reduction in output, others have not deterred incentives to increase production.

Until 1955 the government used quota restrictions to control production and rice farmers were required by law to sell their produce to the government. The quotas were abandoned in 1955, but the government continued to monopolize rice production through contract sales at government established prices. In 1961 the Agricultural Basic Law attempted to shift Japan's excess rural population to urban areas as part of the industrialization and economic development program and to narrow the income gap between farmers and their urban counterparts. Another comprehensive agricultural program in 1968 sought to discourage rice production and instead encourage livestock production and garden farming. In 1969 the government also implemented an acreage diversion program whereby rice farmers were paid to withdraw land from cultivation. In 1971, in an attempt to depoliticize the price-setting system in which farmers had played a large role, the government restricted the quantity of rice farmers could sell the government but allowed farmers to sell their produce through nongovernmental channels.

To some extent these policies have slowed agricultural production by as much as 5.5 percent a year after 1969. This has not relieved Japan of its surplus production problem, however. Since the mid-1960s the government has attempted to balance urban concerns about growing rice reserves, subsidy and storage costs, and export pressures with rural concerns about a declining farm population and growing income gap. Farmers have reacted to attempts to limit production by using smaller machinery and improved mechanization techniques to increase production that is substantially subsidized. In the mid-1960s the government reversed its rice pricing policy. Rice prices were allowed to rise, even sharply, to a level three times greater than world prices. Many consider this move to be the source of Japan's food problem—continued inefficient production supported by costly government subsidies that totaled $3.1 billion in 1975.[16]

To compound the problem, in the early 1970s the yen was revalued upward against the dollar. Higher per capita incomes and changing tastes increased demands for higher protein content and processed foods and also stimualted meat and livestock industries. Japan does not produce enough livestock grains and has since doubled imports of wheat and other feed grains. These imports, until recently, dampened domestic production of wheat and coarse grains, in effect, leaving Japan more dependent then it desires on international markets for food supplies.

These conditions then highlight some of the important structural changes that have occurred in Japan. Although Japanese people are some of the best-fed in the world today, Japan's food policy reflects a situation where farmers and urban workers are caught in a squeeze between rather good structural capabilities and political processes, which will be discussed below.

THE POLITICAL ARENA: THE POLITICS OF RICE PRICE SETTING

A number of conditions in Japan's political arena[17] have contributed to a change in the government's attitude about farm support through larger subsidies. The first is the increasing influence of farmers on the setting of rice price support levels. Rice price setting in Japan is a unique policymaking process in which participants include high government officials, ministers, private organizations, and farmers themselves. Prices are decided outside the normal budget decision process and are not susceptible to the normal constraints that affect other policy decisions. The government's Rice Council allows other advisory committees and executive committees to play a role in deciding prices. Farm groups lobby these government bodies. A National Association of Agriculture Cooperatives horizontally unifies three economic federations, a bank, and a Central Union of Agriculture Cooperatives with a host of vertically unified agricultural organizations at the prefectural and local levels. The National Association claims to represent every farm family and performs the function of a grass-roots organization that continually has Liberal Democratic Party (LDP) support.

A second source of change is the LDP's majority party position in the Diet and control over the prime ministership, which have been increasingly threatened because of declining electoral support. Eighty percent of LDP Diet members are from rural districts, and the farmers, who are becoming less numerous, older, and more conservative, find no other party particularly attractive. Because there has been no significant change in electoral district apportionment over the years, the LDP is more cognizant that its base of support lies in rural districts and may deteriorate without the support of the farmers. Therefore, the LDP has given in to farmer demands for higher price supports.

The third reason for new support policies is directly related to the tightening relationship between Japan's dependence on raw material and energy resource imports and manufactured trade exports that generate economic growth. This has created both a security concern and sensitivity to Japan's growing vulnerability to fluctuating international economic conditions. The Japanese hope that continued rice production will enhance self-sufficiency and that rice stockpiling and higher prices will stimulate production of other commodities and will help keep food prices down for consumers.

Not only is Japan sensitive to evolving political and economic interdependencies with other advanced nations, but it has also been forced to nurture a relationship with the rest of Asia that now accounts for at least one-third of its foreign trade.[18] Continued agriculture production will serve as exchange for the imports on which Japan is so dependent, many of which come from other Asian countries. This explains why American and West European efforts to bring down Japanese trade barriers and hopes for a change based on internal dissatisfaction with price supports have not been forthcoming.[19]

In conclusion, it may be said that an expanding industrial sector absorbed Japan's rural population and served as a market for agricultural products. In turn Japan's industrial sector successfully produced new consumer and technological products that mechanized farm production and generated food surpluses. At the same time, Japan's structural and political conditions since the Meiji restoration have resulted in a significant transformation of Japanese agriculture to the point of inefficient production at taxpayer expense. In many ways Japan's current food problem typifies that of many other advanced nations, especially pluralist democracies where government support of indigenous food products is continued and protected in trade markets even though surpluses accumulate and consumers must bear the burden of high prices and tax supports.

Japan demonstrates the case that countries with self-sufficient food production capabilities will absorb the cost of overproduction in proportion to the amount of decentralized policymaking and weight given to even greater security concerns. A few small farmers survive, not because they are efficient producers given the size of plot they farm, but because they are disproportionately influential (given their numbers) on the outcome of farm support prices. Eighty-five percent of Japanese farmers are only part-time farmers and must depend on government price support or other incomes. More farm labor could be used in urban areas but the trade-off is growth and congestion in already overpopulated cities. Thus, the government now encourages small industry and manufacturing activity in the countryside.

Growing interdependencies have unexpectedly caused Japanese policymakers to support higher agricultural subsidies in order to maintain production levels and enhance national self-sufficiency while safeguarding their own political position. Therefore, we can expect Japn to absorb the cost of wheat and cattle grown at three times the cost of American production and to maintain import restrictions on such products as eggs, fruits, and especially rice. For the time being, Japan has reconciled the issue of inefficient food production with new international political economic realities.

The Centrally Planned Economies

Today, the centrally planned economies of East and Southeast Asia face some of the world's most drastic conditions of hunger and starvation due to the persistence of war and revolution in Laos, Vietnam, Cambodia, and, until recently, in China. Hunger, related to poor growing conditions, adverse weather, a lack of necessary agri-inputs and capital, and poor distribution systems, continues to plague these nations. This section discusses the food systems of a number of centrally planned economies, focusing special attention on China.

In all these countries agriculture is the leading sector in output percentage of GDP and export goods and earnings. At times, China exports rice and a variety of other food commodities and agricultural products to earn foreign exchange.[20] Before recent American military involvement in the area, Cambodia and Vietnam exported natural rubber, the demand for which has since declined due to a change in the terms of trade, caused by the development of synthetic substitutes in industrialized nations, and due to the destruction and deterioration of rubber lands.

China and North Korea show the most significant change in rural population shift to the urban sector as part of industrialization efforts, while in other nations the shift has not been as great (see Table 6.4). Based on a socialist model, government attention has been primarily on their agricultural sectors, especially increasing the output and/or distribution of staple food crops like rice. In some countries wheat and a variety of coarse grains are also grown. Production has increased significantly at times, due to the adoption of better seed varieties, extended land use, improved irrigation techniques, better production methods and marketing techniques.

However, in all these nations food production remains hostage to a number of conditions including unpredictable weather patterns such as monsoon rains and flooding followed by periods of drought in some areas, tremendous population growth (see Table 6.3), and economic and political bottlenecks that limit the range of policy options available to government leaders. Only China has been relatively successful at implementing a population control program. Compounding these problems, these nations (with the exception of North Korea) all face food distribution problems caused by topographical barriers and a lack of transportation networks. Laos is landlocked making it difficult to get its export goods to ports. Laos and China have found it necessary to emphasize rural self-sufficiency and to import food commodities into their major cities. For all the centrally planned economies it is difficult for rural surpluses to reach urban areas to be converted into economic investment resources.

The most significant political and economic bottleneck in Laos, Cambodia, and Vietnam is continued warfare within and among each other. The end of American military involvement in the region has not ended neocolonial linkages to stronger powers like the PRC and the Soviet Union, who play a significant role in the character of regional conditions. With at least two million refugees starving to death in Cambodia alone,[21] it is evident that the food policy used by both the Khmer Rouge and the Pol Pot regime is one of starvation to gain control of Cambodia.

CHINA

Principal agriculture commodities in China are rice, wheat, maize, some other grains, sweet potatoes, cotton, tobacco, tea, and hogs. China is noted for experiments with multiple cropping—raising two crops a year in the north and three in the south—using earlier ripening seeds that make possible the use of 60% more land for wheat productions, and for harvesting two rice crops after the winter wheat crop.[22] The PRC has also been praised for efforts to limit its population growth. The birthrate has been cut significantly to 22 births per thousand.[23] A popular assumption about China, therefore, is that its food production has kept up with its population growth.[24] Another assumption complementary to China is that its model of development, which stresses self-sufficiency and an egalitarian distribution system, would be appropriate for other LDCs, especially the poorer ones with few natural resources and an excess of population.

This section looks at China's agricultural experiment from the structural-political viewpoint and argues that China's agricultural sector is a "food success," albeit a qualified one. Food production problems have been related to the difficulty of maintaining per capita consumption levels commensurate with population growth rates, unpredictable weather conditions, poor transport systems that geographically remove urban centers from agricultural production areas, and relatively unstable political conditions. Whether China's new modernization objectives are reconcilable with efforts to maintain some semblance of an egalitarian distribution system is a question the leadership currently faces.

STRUCTURAL CONDITIONS

Precise statistical information about the severity of hunger and malnutrition in China is not available. However, a number of analysts have tried to determine how much hunger existed in China over the past forty-five years. One recent study suggests that there is good reason to assume that China has not succeeded in eliminating hunger, as is so often claimed.[25] A significant indication of this is the inequality of income distribution between urban and rural areas, between regions, and within regions. The urban populace averages 2,400 calories a day compared to 2,000 a day in rural areas. Urban dwellers also eat 20% more food that is higher in protein content, including twice as much pork. Another study looks closely at grain production trends since 1949 and reveals only gradual production gains and occasional dramatic shortfalls. Robert Field and James Kilpatrick's estimate of China's grain output (see Fig. 6.1 below) is a truer picture of production trends between 1952 and 1976.[26]

Shortly after the Communist takeover, food production recovered rather dramatically. Between 1953 and 1957 production increased an average of 4.2 percent per year. During this time an attempt was made to adopt the Soviet model of agriculture production to China,[27] but this program was discontinued in 1958 because its emphasis on heavy machinery was inappropriate to indigenous conditions and production fell sharply. The growing rift in Sino-Soviet relations also figured in the Chinese decision.

From 1957 to 1960, during the Great Leap Forward (GLF) program, grain production continued to fall due to two years of drought and flooding, communal mismanagement problems, and cutbacks in mechanized inputs into food production. 1961 and 1962 were watershed years when, because of the earlier shortfalls, agriculture finally became a national priority. After Mao adopted the Agriculture First policy in 1961, production rates jumped to 6.6 percent because of increased investment in agriculture and the availability of more "intermediate" modern inputs into the production process.[28] From 1961 to 1966, industry and technology, much of it imported, supplied agriculture with mechanical irrigation systems, new fertilizers and fertilizer plants, new high-yielding rice and wheat varieties, and new types of farm machinery.

However, the results of the Cultural Revolution from 1966 to 1968 were similar to those of the GLF program—another food production disaster due to social disruptions and cutbacks in agriculture research and technology imports. The Chinese slowed their imports of complete plants down to none imported in 1969 and 1970.[29] In the early 1970s production recovered again in response to new technology imports and the use of mechanized inputs in the agricultural sector. But it soon drooped off due to continued high

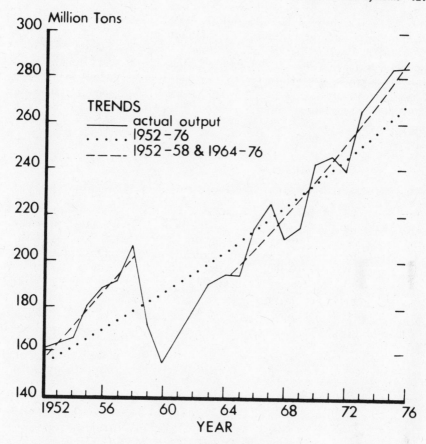

Figure 6.1   Estimate of China's Grain Output
Source: Robert Fields and James Kilpatrick, "Chinese Grain Production—An Interpretation of the Data," *China Quarterly* 74 (June 1978), p. 377.

population growth rates that absorbed surplus production, cutting back on surplus availability for investment purposes. The world food crisis of 1972–74 also cut the availability of food imports to China. At the same time oil prices nearly tripled the cost of fertilizer, and China accumulated large balance-of-payment deficits.

From 1976 to 1978 production leveled off, this time because of poor weather conditions. A number of regions geographically removed from production centers were severely affected, necessitating food imports into port cities also located away from production areas. Since 1961 China has consistently imported an average of 5 million metric tons of grain per year. But in 1976 imports were cut to 2.1 million tons in order to eliminate hard currency trade deficits of nearly $2.5 billion.[30] Although this makes more funds available for technological imports, it contributes to short-run hunger and coincides with recent reports of food shortages.

These production trends and structural conditions would lead us to conclude so far that since 1949 food production increases in China have only roughly paralleled population growth rates. Given the difficulty of sustaining production in the face of unpredictable weather conditions, social disruptions, and international conditions from which China has not been able to insulate itself, one would suspect that claims about China's success as an alternative development model either are exaggerated or need significant qualification.

THE POLITICAL ARENA: OSCILLATING FOOD AND DEVELOPMENT STRATEGIES

As in the case of Japan, in China there is a close relationship between structural conditions and activities in the political arena. Food production and distribution in China has often been drastically accelerated and slowed in response to four political conditions that correspond roughly to the production trends in the periods delineated above. They are (1) polemical leadership disputes about the use of mechanized inputs and development of the agricultural sector versus primary attention to industrial development, (2) the use of imported technology to increase both agriculture and industrial production to achieve for peasants income parity with urban areas or benefit urban industrial center interests, and implicitly fewer people, (3) government centralization-decentralization efforts, and (4) Beijing's role in manipulating farm and industrial prices in an effort to increase food production or ensure distribution.

Even before the revolution ended a leadership disagreement emerged over the timing of land reform and preservation of remnants of the landlord peasant economy.[31] Liu Shaoqui favored material incentives, the mechanization of agriculture, and the use of foreign technology to increase food production and to precede planned collectivization. On the other hand, Mao favored industrial growth in support of creating a proletarian dictatorship via an Industry First policy that would make China a world military and political power. The recovery of food production after the takeover enabled food supplies to continue into cities where resistance to Mao was greatest and made it unnecessary for him to focus primarily on increasing food production. Mao labeled Liu's position ideologically "rightist" but, in keeping with his own goal of modernizing China, allowed some peasants to keep landholdings. According to Mao's view, agrarian reform, collectivization, and the mechanization of agriculture would take place gradually, commensurate with conditions in industry.

Thus, after the revolution production techniques centered around traditional methods and agricultural inputs. Agriculture output was controlled by a combination of public and private trading enterprises that taxed a percentage of production. The remainder was sold to government trading companies or private traders or at peasant fairs. The state sought to gradually control private traders by setting advance purchase prices, which, it hoped, would stimulate production and improve farm incomes.

From 1953 to 1957 government efforts at centralization were enhanced by attempts to impose a Soviet model of agricultural development on China. Labor exchange groups and mutual aid teams were formed and through state owned Machine Tractor Stations (MTSs) the government centralized farm policy planning and implementation.[32] When the limits of the Soviet model were realized Mao pressed for the continuation of another form of collectivization. He favored cooperatives before the mechanization of agriculture. Peasants were organized into some 26,000 communes averaging 155 families in each. This represented a shift in Mao's thinking about the value of emphasizing agricultural as much as industrial development as well as the success of his new ideas over protests from within the Party (e.g., by Liu Shaoqui). This shift also coincided with Mao's fear about the reemergence of capitalism in the countryside. Politically, he feared a loss of control over integrating the communal structure and thus a loss of his base of support in the rural peasantry.

Therefore, from 1957 to 1961 Mao attempted to revitalize agriculture and decentralize policymaking via the GLF program. The MTSs (viewed as bureaucratic elitist) were dismantled and the ownership of machinery decentralized in the new communes. Attention turned to the use of semiindustrialized medium-sized tools built by local industries that served to link agricultural mechanization to national industrial development and to overcome China's bad transport system. The effect of Mao's policy resulted in a significant increase of labor in agriculture and the continued organization of the communal structure. Compulsory production purchase quotas were introduced at fixed prices based on average yields. Peasants could keep 5 to 10 percent of their produce. But rural markets and peasant fairs were discontinued in an effort to emphasize ideological incentives and communal production and marketing activities.[33] However, production did not increase and it became clear that Mao "was ready to accept shortcomings in production as long as collectivization enabled the effective mobilization of the rural masses."[34]

Politically, the Agriculture First strategy of 1961 was an attempt to quell criticism of the GLF program, but at the same time it represented a bureaucratic effort to recentralize control of policymaking over entrenched decentralized forces. Actually, the decentralization of communes in 1962 that made production teams the basic accounting unit to determine production quotas represented a centralization of the government's role in controlling China's intersectoral exchange process. Communal farm teams could now decide on production plans formulated by state planning commissions, but the two groups together would decide output targets, how much would be sold to the state, and how much could be kept as a ration.[35]

During this time food production increased; so did Mao's concerns about the growing cleavage between the urban and rural sectors and the effect of mechanization on the two economies. His plan of "Agriculture as the Foundation and Industry as the Leading Factor"—the "Walk on Two Legs" strategy—was actually a political compromise in order to avoid developing industry at the expense of agriculture. As the urban-industrial sector grew, Mao's influence began to wane, and supporters of the Industry First strategy had more influence on policy planning and the decision-making process. Thus, Mao agreed to private plots, family sideline activities, and payments made according to labor. Cities still needed food supplies if industry were to supply machinery to rural areas and absorb displaced labor.

Before the Cultural Revolution, rural industry had grown because imported technology had been diffused out to the countryside.[36] Agricultural output increased and exports were needed less to earn foreign exchange. As was the case of the earlier GLF mass campaigning, Mao again had a firm foundation upon which to assert his ideological and political authority. He could afford to initiate measures that he hoped would make China self-sufficient and independent of external political and economic linkages. But Mao's new decentralization efforts actually enhanced the government's role in production by Red Guard enforcement, down through a more tightly coordinated party and bureaucratic machinery. As before, Mao temporarily prevailed because he gained control over the CCP, which solidified its rural strength enough to control communal activities.

After the Cultural Revolution it became more difficult for Mao to maintain his leadership as a result of entanglements with the growing numbers of pragmatic leaders like Zhou Enlai and Deng Xiaoping. Predictably, the bureaucracy reasserted its control over the economy in the face of a new set of structural conditions related to increasing dependencies on the international system. Since 1971 production has increased but not at a rate acceptable to China's leaders. Except for the short reign of the Gang of Four after Mao's death the government has centralized its power around a new ruling elite that draws its support form a growing bureaucracy and urban-industrial sector. More significant is a change in the view of the leadership itself. With the reinstatement of Vice-President Deng Xiaoping under the moderate Hua Guofeng, leadership policy positions are less ideological, more tolerant of less-than-egalitarian life-styles,[37] and supportive of industrialization and outward-looking trade strategies.

Government centralization efforts have also been furthered in an attempt to ensure the success of the new modernization program, which includes the renovation of agriculture, industry, national defense, and science and education, in order to accomplish three national objectives: (1) increase food production for the growing population, (2) develop an industrial base that can meet consumer demand and help modernize the Chinese army,[38] and (3) achieve for China the status of ‹rst-rank industrial power by the year 2000. Much has been made of China's new modernization program. As one writer put it, China is "shucking Maoist ideologies so quickly these days that it seems to be conceding a colossal blunder."[39] Actually the new outlook signifies a diffusion of the Agriculture First emphasis of the Maoist eras and an optimistic outlook about the role of industrialization and its possibilities for mechanizing the agricultural sector and increasing production.

Today, China's national food policies are made at higher governmental levels, which set the boundaries for more specific policies implemented at lower levels. Annual production quotas and government procurement targets are established by state planning commissions to which brigades and communes are responsible via the vertical chain of command. County planned targets account for government procurements (much of which go to urban areas), reserves, ration amounts, and income payments.

A significant change has occurred in the views of the populace who want a bigger "payoff"—i.e., a higher standard of living—for their efforts over the past thirty years. The government has responded with higher wages in industry, across-the-board wage hikes, bonuses, overtime pay, points adjustments, and work incentives.[40] In the rural sector the desire for a payoff is manifested is a preference for wheat instead of rice or potatoes, which are regarded as "poor man's food." Planting schedules have been adjusted in some cases.[41]

In the past, peasant incentives were often subject to changing political conditions, but the new modernization program seeks to ensure that peasants benefit directly from tangible incentives.[42] Toward this end the government has initiated a number of new measures including higher prices for government-procured vegetables and livestock and reduced grain surplus procurement percentages (down from 90 to 70 percent) sold to the state. They guarantee the right to work private plots and have peasant fairs.[43] Another move is to encourage commune food production and self-sufficiency of certain commodities (especially vegetables) in and near cities.[44] New priorities include irrigation projects in the north and the integration of industrial and commercial operations with commodity production in the same area. More state loans are to be available to low-priority locales. In the area of agriculture inputs the government seeks to better control management, labor, and supply lines in order to enhance economic efficiency. The government has also consolidated and reorganized a number of agriculture-related agencies in order to promote efficient policy planning and implementation.[45]

The most critical factor leading to the adoption of the new modernization strategy has been policymaker acceptance that earlier development objectives and agriculture policies could not achieve modernization via income equalization without more emphasis on growth. Yet the new modernization program is a gamble that attention to economic growth and outward-looking trade strategies will hasten production soon enough and that the amount will be large enough to justify the risk of less emphasis of egalitarian distribution efforts. Agricultural production must increase dramatically if the PRC is to feed its growing population and pay for a modernization program with foreign exchange earned from agriculture exports. Chairman Hua Guofeng also announced at the National People's Congress in March 1979 the goal of increasing grain production by an average 4.3 percent a year or by 400 million tons by 1985.

China's new leaders are also more aware of the reciprocal and beneficial relationship of sectoral linkages. They hope that imported technology will again produce mechanized agri-inputs that will result in increased agricultural output and that industry will be able to absorb displaced labor. They also hope that enough agricultural surpluses will be generated and exported to earn the foreign exchange needed to purchase new technologies.[46] On the other hand, given the inelasticity of demand for food, even when agricultural production increased in the past by the use of mechanized agricultural and technological inputs, the number of hungry people in China may have increased due to the skewed income distribution throughout the PRC that results from an emphasis on stimulating economic growth. Ironically, we would expect to find fewer cases of extreme hunger during decentralization periods like the GLF or the Cultural Revolution when production decreased. In fact, if we are concerned about society's greatest numbers, China's newer leaders may have already overemphasized economic growth and reliance on market distribution measures. Recent cutbacks on spending for the modernization program, a temporary halt to new trade deals, and increases of food imports may be misinterpreted by the international business community as an attempt to slow down and assess the program in its first stage of growth.[47]

These cutbacks may actually signal that the severity of hunger in China is reaching politically unacceptable levels and that China is unprepared as of yet to play as significant a role in world affairs as it would like. Paradoxically, because of the severity of structural conditions, especially hunger, the range of domestic and foreign policy options for China's leaders and their chances for military and political success related to security concerns vis-à-vis Russian hegemony in Asia[48] could be limited. What the Chinese label a "punitive" adventure into Vietnam found them unprepared to punish the Vietnamese as much as they would have liked.

Low-Income Market Economies: A Profile

This section profiles the food system of Thailand, Malaysia, Indonesia, Burma, and the Philippines, where policymakers also face a number of structural constraints, such as poverty and poor transportation networks, that prevent their rural labor force—the majority of the population (see Table 6.2)—from purchasing and recieving adequate nutritional food supplies and effectively linking rural consumption patterns with national development objectives. Some of these nations have increased agricultural output levels substantially along with their economic growth rates. Yet hunger remains a persistent condition that in some cases thwarts the success of development efforts and, in other cases, is even caused by such efforts.

STRUCTURAL CONDITIONS

These five nations are some of the largest in the region yet some of the poorest in the world. One study estimates that 355 million people (nearly 40 percent of the population) live below the poverty line.[49] Subregional food production has averaged 3 percent growth rate per year and has just barely kept pace with a population growth rate of 2.5 percent. This has not prevented malnutrition or even starvation in many areas, such as Indonesia's Floria Island. Per capita incomes are some of the lowest in the world (see Table 6.3) and wide income differentials exist between sectors and also among members of the agricultural sector. The rural population of Indonesia are the "worst off" in terms of production meeting population growth and domestic demand, while the rural population of West Malaysia and Thailand are the "best off."

Per capita calorie intake for the subregion is 2,100 calories per day, well below the daily minimum recommendation for protein and caloric intake (see Table 6.2). Since 1956 only Thailand, Burma, and Malaysia have significantly increased per capita food production and consumption. Therefore, some have had to supplant crop production with food imports, which also makes them susceptible to world market price and supply fluctuations like those of the 1972–74 world food crisis. Indonesia has become the world's largest rice importer (2.6 million metric tons or two-thirds of the world's export for 1977). Because the demand for cereal is virtually price-inelastic and because most rice is locally grown and not traded on the international market, large shortfalls drastically increase food prices and, in effect, severely restrict the nutrient intake of the vast majority of members of these societies.

These nations continue to encounter high rates of population growth and migration to urban areas. A number of birth control programs have been implemented that, because of government support, show signs of having a positive effect on the birthrate. President Suharto's support for the national program apparently contributed to a decline in Indonesia's annual growth rate.[50]

In this subregion food production remains extremely susceptible to weather conditions, which in the early 1970s brought significant droughts, flooding, and typhoons. Over a number of years rice and wheat production increased in response to intensive modern technological inputs successfully diffused in all five nations during the Green Revolution. But rice accounts for anywhere from 65 to 80 percent of caloric intake and the area of land utilized for production has generally not increased. Even so, new inputs of high-yielding rice and wheat varieties are looked to for increasing yields. However, the effect of new technologies has not been neutral, adding to unemployment that at times runs parallel to underemployment problems. The effectiveness and efficiency of new technologies also reached a point of diminishing returns due to the high cost of oil and agricultural inputs in the mid-1970s. Despite real improvements in production methods then, consumption conditions are judged to be not much better or even worse today than they were ten years ago.

POLITICAL ARENAS

Three controversial issues for policymakers in these nations are food-pricing policies, land reform measures, and trade, aid, and foreign investment policies. Generally, these governments share a number of common characteristics involving the regulation of food prices and market activity in which they all participate more actively due to changes in political and economic opportunities and structural conditions. Throughout the 1950s and 1960s, their plans gave priority to increasing industrial and manufacturing output in urban areas. Pricing policies were urban-biased; they were aimed at keeping food prices low in cities, the location of most of the regime's political support. Farm support measures were hard to manage alongside cheap food programs for the urban populace. Governments cut financial assistance to and investment in the agricultural sector because of a general downturn in domestic and international economic conditions (namely a lack of revenue earned from export sales, inflation, the high cost of living).

Economic-industrial growth strategies proved to be inappropriate to indigenous conditions. Greater numbers of smaller farmers demanded higher prices and incomes, and because of the political and social upheaval that often accompanies such demands, many governments switched to support of the agricultural sector with higher prices, which they believed would stimulate the adoption of new technologies, encourage the use of fertilizers, and increase agricultural output, especially the commercial crop variety that earns foreign exchange. Pricing policies then were generally used to enhance food self-sufficiency and at the same time "minimize foreign leverage and the risk associated with dependence on international markets for food supply."[51] Table 6.7 highlights types of price systems and prices.

As is often the case, and as may soon happen in China, these nations became caught between a "rock and a hard place." Although intentions were to help farmers, the unintended consequence of high food prices hurt the lowest income groups the most. Over the long run, high prices reached a point of diminishing political, social, and economic return and in some cases discouraged investment in better irrigation or technological inputs. At times they were too rigid to allow for needed adjustments as production levels changed. Governments also found it difficult to stabilize grain prices. The pattern of farm organization follows the pattern of land utilization in Asia. "Large commercial farms prevail in areas of plantation crops, while small

Table 6.7 Procurement or Support Prices for Rice in Low-Income Market Economies

Country	Product	Type of price system	1969-70	1975-76
			($/ton)	
Burma	paddy	procurement price	36	56
Indonesia	paddy	support price	50	101
Malaysia	paddy	procurement price	103	168-187
Philippines	paddy	support price	93-104	166
Thailand	paddy	average wholesale price of paddy No. 1 in Bangkok	55	117

Source: FAO, The State of Food and Agriculture, 1978 (Rome).

farms are practically the sole units of the region's output of traditional crops."[52] Thus, large commercial farmers benefited the most from high food prices. Not only did they receive better terms of credit and hold control over technological inputs, but high prices guaranteed producers of marketable surpluses a profit while small farmers and the poor had to spend a greater proportion of their incomes for food and production inputs.

Land tenure systems are another problem that cause much hunger in this region because they effect the distribution of income. Tenure systems are quite diverse, and it is hard to say what type of system is the most productive for each nation.

Many Southeast Asian nations adopted land reform programs, but their success was often prevented by a lack of government commitment to reform, which reflects the distribution of political power related to tenure systems. Land reform programs were quite often political footballs used to attain sociopolitical objectives and were not necessarily designed to reform the tenure system. The case of the Philippines demonstrates this point, when an American-initiated land reform program failed because of a lack of government support.[53] Between 1946 and 1972 the government had backed reform efforts, but only as part of an attempt to quell insurgents. The program itself was not centralized enough to be effective. Large plantations that specialized in export crops like sugar and coconut were exempt from reform while lands that produced rice and maize were reformed, reflecting the government's prolandlord and private industry outlook. With the introduction of martial law in 1971, President Marcos became a more forceful advocate of reform due to his decreased dependence on political parties or landlords. The case of land reform in Indonesia in 1971 also demonstrates the case that land reform programs not accompanied by infrastructural changes resulted in productivity benefits going to the middle and large farmers because institutional arrangements favored them.[54]

Trade, aid, and investment policies of these nations have had a critical impact on their rate and pattern of economic development. Exports of food and nonfood crops not only earn foreign exchange to purchase needed imports and large quantities of food but also serve as employment industries. A high percentage of exports (especially in Malaysia, Thailand, and the Philippines) have been food commodities, which are generally declining in their value of total exports. Although world trade in agricultural commodities has increased in the last fifteen years, the share of product trade for nations of this subregion has declined substantially. Factors that account for this are both domestic and foreign. Rice surpluses in Thailand and Burma have been consumed by population growth, and a shift in the comparative advantage of food commodities produced by these nations has gone to the advanced nations. Terms of trade for tea, rubber, and palm oil in particular have worsened because of lower demand and a near saturation of these products in the developed nations and because of their protective tariffs and import restrictions. Therefore, many of the poorer nations sought to diversify their export products and to move into the food processing business instead of relying on "raw" sales, cutting dependency on traded agriculture commodities to between 4 and 37 percent of exports. The Philippines,

Malaysia, and Thailand have rapidly increased the proportion of manufactured goods they export. Malaysia has benefited from higher rubber prices and Thailand from earnings for rice, textiles, and tin.[55]

Yet the continuing need for expensive fertilizer, technology, and food imports, as well as a lack of export earnings accompanied by inflation in the advanced countries, has left many of these nations with large balance-of-payment deficits.[56] There has also been a general decrease in the amount of bilateral concessional aid from OECD members to these nations, replaced by other international sources, including OPEC assistance to Indonesia and Malaysia.

During the 1972–74 world food crisis, which coincided with OPEC oil price hikes, many developing nations learned that "the world market system could not, in the future, be relied upon to provide needed goods and services in sufficient quantities at reasonable prices."[57] This has forced them to focus more attention on self-reliance, as distinguished from self-sufficiency, and producing needed commodities themselves in an attempt to blend the objective of economic growth with a better distribution of commodities to low-income groups. Many of these poorer people are members of ethnic groups that governments or leaders use food policies to help or hinder.[58]

The range of options for policymakers in low-income market economies continues to narrow considerably due to a lack of insulation from fluctuations in international market conditions. There has been only a slight weakening in the dependency of these nations on developed markets. They are still dependent on trade, aid, and/or international organizations and finance institutions to fill income gaps in their investment strategies. Popular government leaders and military organizations play a critical role in formulating policies that are likely to reflect an attempt to enhance support for their own political position. This was demonstrated in Indonesia, where Suharto gave the national oil company Pertamina a free hand and based national development plans around export oil earnings.

In conclusion of this section, it can be said that the low-income market economies are far from realizing their development objectives. What these Asian nations have in common with most of their neighbors is the constraints they face in the form of cultural traditions, structural conditions including population pressures and fluctuating agricultural output trends, and remnants of colonial infrastructures that exist side by side with traditional institutions. Their Japanese and Chinese counterparts have more developed administrative and infrastructural institutions and a clearer conceptualization (even if widespread agreement or support does not always exist) regarding the agriculture system and government's relation to it.

Hunger and National Asian Food Systems: An Assessment

Given the characteristics of food systems in these three types of economies and after closer inspection of the two most popular development strategies in the region, two important issues left to be dealt with are, first, whether Japan or China's development strategies are appropriate and adaptable to the poorer economies in the region, and second, if they are not, what kind of strategy will enhance development and account for society's "bottom 40 percent"?

Japan and other relatively better-off high- to middle-income market economies do not face major hunger problems. They have controlled the timing of their intersectoral exchange process so as to defuse its detrimental effects. Few constraining structural conditions leave the boundaries of political arenas flexible enough so that policymakers can choose the appropriate technology, and significantly, the most appropriate location-specific development strategy.

Japan's "induced" development strategy (i.e., a derivation of the ideal type of Western model) used outward-looking trade strategies and imported technology to fashion mechanized inputs that increased agricultural production and fed a growing population, complementing the development of Japan's industrial sector.

These countries regulate food prices and market conditions in conjunction with sustaining economic growth and development objectives. Japan and Taiwan even provide displaced labor with alternative sources of income. Because of their economic efficiency and active trade status, the more developed Asian countries encourage food production to the extent that they desire to be self-sufficient producers. They import nearly all the food required by the national and per capita income demand. What remains of their agricultural sectors will probably produce food at prices well above international market levels.

Beginning with the Meiji, strict control of food prices was avoided in order to stimulate food production and achieve self-sufficiency. But food policies were determined by a highly centralized bureaucracy whose power resided in Tokyo.[59] As in many other advanced industrial nations with decentralized policymaking structures and declining agricultural population, price controls and support measures are now the result of the political influence of remaining farmers. These measures are also adopted because options are more limited by Japan's small geographical size and its vulnerability to international political and economic conditions, which require it to maintain sufficient food production levels while specializing in industrial and manufacturing activities.

Policymakers in the centrally planned and low-income economies face far more serious structural conditions that force them to more consciously consider food production and distribution measures as part of their development or nation-building goals. Structural constraints play a greater role in consideration of food and agriculture policies. At the same time, many of their political arenas have yet to be organized, institutionalized, or accepted as legitimate. Of the centrally planned economies, only China, North Korea, and Laos have solidified control over the policymaking structure and economy enough to make and implement policies aimed at modernizing their society. On the other hand, Vietnam, Cambodia, Indonesia, and others still use hunger and even starvation to eliminate hostile ethnic minorities or to gain control of society.[60]

The intersectoral exchange of the poorer countries has been slowed by a host of structural conditions including uncontrollable population growth rates, weather conditions, and poor transport systems. The political arenas of nearly all developing nations in the region are dominated by struggles between authoritarian leadership structures and growing bureaucracies. In many cases a cult of personality is the determinant factor that conditions food policies and development strategies. China is a classic example of a nation whose rural population served as a political base of support for the leadership. Thus, Mao would naturally oppose population control measures because they enhance a shift of population from rural to urban areas.

However important or desirable were the ideological goals of Mao's brand of socialism, it became evident to China's leaders that modernization would have to include some capitalist activities—that structural limitations require more pragmatic solutions. China's development strategy has not been the radical model so often extolled in textbooks. While ideological goals pushed strategies in one direction, pragmatic problems necessitated practical solutions that pulled them in another direction.

Thus, agriculture production increases are observed from 1949 to 1951, 1960 to 1966, and from 1971 to the present when the government attempted to centralize its power, gave more attention to industrialization, and encouraged trade development. Production decreases occurred from 1956 to 1960 during the GLF campaign and from 1966 to 1968 during the Cultural Revolution when Mao's primary attention focused almost exclusively on agricultural development and self-sufficiency strategies. Controls over food prices were gradually tightened in order to cut food costs to urban dwellers, while incentives for rural laborers oscillated between ideological rewards and more tangible incentives like private plots, higher procurement prices, profit sharing, and peasant markets. During decentralization periods production decreased when prices were lower and collective incentives emphasized, adding to the problem of accumulating surpluses and transporting them to cities. The development of the communal production structure served as an attempt to distribute income more equitably, while self-sufficiency objectives served to offset poor transportation systems and the need for more food in urban areas.

The food policies and development strategies of the low-income market economies mix public and private enterprises with the use of price incentives, the effectiveness of which are location-specific. Some governments (such as Burma) have adopted strategies that theoretically give much attention to social justice, whereas others (such as Indonesia, Malaysia, and Thailand) lean more toward market approaches. Production conditions in these countries remain susceptible to a host of factors that include unpredictable weather conditions, poor transport systems, large populations and high growth rates, lack of credit to small farmers, bureaucracies that cannot effectively implement and evaluate policies and programs, inefficient and socially unsupported land tenure systems, and finally, skewed income distribution, which makes poverty the most critical item on the policy agenda.

On the other hand, the ability of these nations to significantly invest in industrialization, however, is limited by a shift in the terms for their semimanufactured products. This necessitates an emphasis on rural-sector development and national self-reliance via a mixture of import restriction and substitution policies to protect against the disruptions of the developed economies, who can more easily force others to adjust and absorb the effects of their trade practices.

Whether indigenous conditions of other nations can be molded to fit either the Japanese or the Chinese model is doubtful without a more centralized or authoritarian political structure to enforce such an attempt from above. Japan and China's development strategies complemented local political, economic, and social systems. Japan emphasized economic growth and the use of market forces, while China attempted to focus on distribution, which required more government intervention and stricter market regulation. But in many of the poor LDCs the adoption of China's strategy may result in a loss of economic growth. An inspection of China's experiment reveals that popular participation has not increased production enough to protect against inconsistent weather patterns, nor has it yet contributed to effective consumer demand that stimulates enough investment in industrial development programs. The failure of China's socialist experiment has taught it that achieving modernization requires more attention to economic growth and acceptance of something less than egalitarian food and income distribution outcomes. Yet adoption of Japan's more open market strategies could result in skewed income distribution and cause more hunger. Japan is unique in that it has been able to establish adequate consumption levels as a result of the "trickle down" plan of economic growth.

It would seem, then, that the success of food policies and development strategies of any model depends on each nation's uniquely arranged indigenous structural features, political arenas, national objectives, and relationship to other nations in the region. This makes one pessimistic about the chances of constructing an appropriate strategy for Asian economic development—one that incorporates policies that contribute to a solution of food production, distribution, and consumption problems in the region. However, the question remains whether the poorer nations in the region can ever get out of their food production-distribution-development quandary.

The only alternative to these models may be a modified growth model that emphasizes distribution almost as much as, but not more than, growth. As this work has shown, these goals are not necessarily complementary, nor are they always mutually exclusive. A number of developmental economists have revised the traditional economic growth strategy to include requirements that governments ensure food distribution to society's bottom 40 percent. In practical terms I prefer the "self-reliance" strategy to ultimately solve hunger problems. Whereas self-sufficiency objectives emphasize autarky and usually employ import substitution measures to increase the production of consumer goods, the self-reliance strategy incorporates a goal of rural development within it. Depending on a nation's comparative advantage, agriculture receives as much attention as industrial development.

At least the interrelatedness of the two sectors in consciously considered. According to the self-reliance strategy, industrial development may not be the goal for LDCs to pursue. Economic growth may result from an emphasis on tertiary sector activities.[61]

This strategy should incorporate variance in national systems capability to efficiently utilize resources. In the case of land tenure systems Hayami and Ruttan suggest not only that there is no single optimum land tenure system, but that tenure is location-specific.[62] For many of these nations, small land holdings that are more labor-intensive seem to work best.[63] What is not clear yet is how the size of holdings, new technological inputs, and increased production are related. The key to land reform success seems to be, first, a *secure* tenure system, and second, a complementary institutional setting that serves the interest of the small farmer and not just of the rural elite. Over the past decade, some land reform programs in Southeast Asian nations have been strengthened. But land reform alone accomplished little. Landlords are a political and economic base of power for leaders. But national leaders can end run their influence by supporting the creation of employment opportunities in urban-industrial and countryside projects that absorb labor.

Another key issue is food prices. High food prices can mean the difference between success in surplus accumulation and investment in both public and private agriculture, and the need to immediately feed the hungry. Procurements can be used to build up stocks to support urban consumption. The size of stockpiles should be intentionally controlled (not necessarily kept small) to prevent wide price fluctuations. Burma and Thailand have tried to build their own reserves in an attempt to regulate imports and exports and to stabilize domestic rice markets. They also participate in the ASEAN food reserve program (discussed below), which draws on national surpluses. Indonesia and the Philippines have spent more for public and private agricultural facilities and boosted support for small farm lending institutions.

The important point, however, is that the chances for balanced growth and hunger relief in the poorer nations are worsened unless more consciously applied distribution programs can be incorporated into development strategies until such time as minimum food consumption requirements are met. Thereafter the exact ratio of attention to growth and distribution strategies is proportional to politically acceptable losses of either economic growth or capacity to absorb the cost of poverty and hunger.

A Regional Structure and Approach to Asian Food Problems

This concluding section deals with the development of an institutionalized Asian regional organization and prospects for its contribution to the alleviation of some of the region's most drastic hunger problems. As in the case of the EC, the stimulus for Asian integration has been attempts to deal with security concerns, whereas a more integrated regional structure that manifests actual policymaking authority has been slow in coming.[64] But there is now good reason to believe that closer cooperation between members of the Association of Southeast Asian Nations (ASEAN) will both indirectly and directly contribute to economic development and solutions to regional food problems.

After World War II the United Nations Economic and Social Commission for Asia and the Pacific (ESCAP) and a number of private and public regional institutions began to give Asia a regional identity and to foster regional cooperation in diverse fields."[65] The most significant of these has been the Asian Development Bank, a Reserve Bank and Development Fund, the Asian Industrial Development Council, a number of Commodity Communities (including rice and tin), and a host of functional projects like Trade Expansion, Monetary Cooperation, the Mekong Basin Project, and the Asian Highway and Trans-Asian Railway. However, the subregional organization with the best chance of institutionalizing a regional structure that can significantly benefit Asia is ASEAN, whose membership includes one high- to middle-income country (Singapore) and four low-income countries (Indonesia, Malaysia, Thailand, and the Philippines). Organizers hope that ASEAN will be "the indigenous machinery whereby intraregional conflicts could be minimized if not settled."[66]

In 1966 ASEAN evolved out of two previous unsuccessful attempts at subregional organization. From its inception, security issues have been high on ASEAN's agenda.[67] The original motivation behind the formation of ASEAN came from a unified effort to minimize feelings of insecurity and enhance national independence behind Indonesian president Suharto's efforts to create a nonaligned Southeast Asian regional organization independent of great power domination. In the late 1960s the organization focused on the PRC's efforts to expand its sphere of influence in Southeast Asia.

Since U.S.-PRC rapprochement in the late 1970s, attention has gradually shifted to other security concerns. The first is recent Soviet efforts to surround China with a set of client states more dependent on, and influenced by, Moscow than Beijing. ASEAN members have gradually improved relations with the PRC, and some with the Soviet Union. Still, caution about superpower intentions prevails among the membership, especially in the face of hostilities between Vietnam, Cambodia, and Thailand. External threats have "tightened ASEAN's cohesion, helping it to make progress toward becoming a true political and economic community."[68] The continuing support of local communist parties by China or the Soviet Union has also caused members to seek and form new bilateral security agreements as well as arms trade and standardization agreements.

A relatively new and important goal on ASEAN's agenda is economic cooperation in regional development, which has not been as easy to accomplish as cooperation for the sake of security. At a number of conferences members have reached agreement on a variety of economic issues, including tariff reductions, preferential trade agreements, and most significantly, a surplus rice stockpiling and distribution plan (discussed below). In 1976 ASEAN members studied the feasibility of allocation in "package treatment" form their fertilizer and iron and steel industries.[69]

Intraregional trade had not increased as much as was hoped. It has been hindered by continuing prewar bilateral colonial relationships or the lack of goods to trade. In the 1970s intraregional trade consistently accounted for roughly 15 percent of ASEAN's total trade. Thailand's rice exports and Singapore's commercial activities accounted for the largest share of trade. On the other hand, extraregional trade has increased between some nations, especially Singapore and other advanced industrialized countries. ASEAN's comparative advantage still lies with petroleum from Indonesia and petroleum products in Singapore. The entire region also has a number of other raw materials and minerals that can be used to promote industrial development: iron in the Philippines, tin and iron in Malaysia, and coal in Indonesia.

The important issue as far as food and development policies are concerned is to what extent growing cooperation or the ASEAN organization can help members or the larger region with what seem like unsurmountable hunger problems. At least 75 percent of the population of ASEAN countries are engaged in agricultural production activities. As noted above (see Table 6.5), the contribution of agriculture to GDP in these countries has declined significantly over the past two decades. However, this trend does not represent any real improvement in the problems associated with a "backward" agricultural sector—extreme poverty, overpopulation, unemployment and underemployment, and low productivity. As discussed above, through the mid-1970s rice production in ASEAN countries did not significantly increase.

Yet close cooperation between members on issues of economics and security first of all establishes an underpinning for institutional-infrastructural reforms. Many rural elites fear revolutionary change and resultant instability. Thus, for example, in the case of land reform, it appears they "might agree to an orderly land reform . . . having fully realized the old maxim that it is better to give up something in order not to lose all in a total revolution."[70] This tendency is supported by growing urban interests that seek to maintain access to food supplies and thus are likely to support peasants over landlords. Secondly, the ASEAN organization specifically contributes to "balancing" the development of agriculture with industry and tertiary activities by supporting the view that agriculture is a dynamic part of economic development. ASEAN is now beginning to pay more attention to such items as the availability of rural resources, agricultural investment, rural development programmes, improved production technologies, marketing, and credit systems.

ASEAN's role in the economic development of this region is significant in terms of encouraging the adoption and implementation of appropriate strategies for national and regional development. These strategies will have to be more "integrative" and "multidimensional" than previous strategies, which will necessitate more political will and intervention on the part of governments in order to accomplish their objectives. This the ASEAN nations seem more willing and capable of doing.

ASEAN's most immediate impact on hunger conditions is most likely to be its direct relief measures and its newly adopted common agricultural policy, which includes a plan to build up reserve rice stockpiles. Each ASEAN member contributes a percentage of its rice surpluses (which

Singapore has to import) to the reserve, to be distributed according to need during times of shortages. Common agricultural policies include measures on agricultural research and technology, training and extension services, and a plant and animal quarantine program.[71] Even though four of the five ASEAN nations export a high percentage of agricultural commodities, ASEAN hopes to capitalize on Asia's great rice-producing potential. If rice production can be developed and supported by "regional" policies so that surpluses could continue to be stored without great cost to members, rice exports may give Asia a comparative advantage that will enable it to compete with other grain-exporting regions. Thus, rice accumulation serves three useful purposes. Besides building up surpluses for emergencies, they contribute to market price stability, which theoretically should help keep prices from rising and thereby directly benefit the poor. Reserves also act as a subsidy in that they provide farmers with a guaranteed market, which ASEAN hopes will enhance self-sufficiency and "result in the more effective utilization of agricultural resources."[72]

Despite this positive outlook, ASEAN's further institutionalization has been slowed by a number of conditions[73] including strong feelings of nationalism on the part of members, which impedes long-range planning and prevents further cooperation. Another problem is the competitive agriculturally based economic structures of four of the five members, who continue to protect what successful industries they have been able to develop. Third is the continuing heterogeneity of national social systems. Continuing traditions, ethnic, religious, and cultural animosities plague efforts to limit separatist tendencies.[74] Finally, as a result of the as yet decentralized structure of ASEAN itself, the organization is accused of lacking "organizational muscle" and political clout.

But these problems should not detract from the integrative successes accomplished by ASEAN so far. One FAO official I spoke with from this region firmly believes that "as ASEAN gets stronger, politics takes care of itself." Other national and regional factors seem to compel nations to cooperate even more. East and Southeast Asia is the most dynamic political and economic region in the world. Japan and China are the region's most influential actors and, because of their new superpower status and foreign economic policies, they are likely to seek development of a positive relationship with an increasingly more viable ASEAN organization. As dependency and vulnerability considerations take precedence over internal pressures to insulate Japan from the rest of the world, Japan is compelled to search for continuing sources of energy and raw materials to feed its "incorporated" industrial structure. Thus, Japan has invested significant amounts of capital in Southeast Asia, especially in Indonesia and Malaysia. ASEAN has helped to secure investments in the region in exchange for trade barrier reductions against ASEAN products.

The PRC is also looking to Japan, Western Europe, and North America to help with its modernization program. The PRC has recently made a number of deals that include direct foreign assistance loans from Japan and Singapore to finance grain trade with Argentina, Australia, and Canada over a three-year period.[75] Given China's drive for modernization and rapprochement with the West and with ASEAN, we are likely to see this emerging relationship play a fundamental role in the development of the region.

Other major invesments in this subregion have come from Western Europe, and have been rather successful for ASEAN. The United States also invests heavily in the Philippines and in Thailand. ASEAN's support of outward-looking trade strategies has attracted some investment in local industries and commercial activities. Given the continuation of slowed economic growth in the Western world, ASEAN is likely to have a significant impact on regional political economic conditions in Southeast Asia.

One would like to conclude this chapter on an optimistic note. We can say that, due to a number of recent developments, linkages between Asian nations are increasing. We can look for ASEAN to gradually help relieve hunger problems in the region. The more cooperative members are, the more likely they are to create an environment conducive to generating economic growth. Regional cooperation and a more institutionalized regional organization that fosters economic integration can, to some extent, level, or at least serve to help close, the income gap among ASEAN members and nonmembers alike.

But one must be realistic about Asia, given its history and the current problems that defy easy solutions. It would be presumptuous at this time to say that ASEAN now has, or even will have, the same positive effect on economic development in Asia that the EC had on economic recovery and growth in Europe after World War II. The history of development of the advanced industrialized and more successful developing nations in Asia may not predetermine the fate of the poorer Asian countries. Asia is developing a regional perspective and a viable regional organization, but overwhelming structural conditions and problems that must be dealt with in a variety of political arenas considerably narrow the range of policy options available to national leaders. Unless more appropriate and specific national food and development strategies are forthcoming—strategies that are also at the same time not detrimental to regional cooperative efforts—there will be more hunger and even starvation in a region where there is no obvious model solution to the region's most pressing problem.

Notes

1. Many developmental economists posit that the timing of the "intersectoral exchange" or structural shift of rural labor and population to urban areas corresponds with a shift from an emphasis on agriculture to industrial-manufacturing and tertiary sector industries. As an evolutionary modernization trend other developed nations have gone through, developing nations will follow to the extent that their comparative advantage in nonagricultural industries is realized. What I refer to in this paper as an "intersectoral exchange" between the rural and urban sectors Bruce Johnston calls a "structural transformation" of the two sectors. He suggests that this condition is almost a law. See Bruce Johnston and Peter Kilby, *Agriculture and Structural*

Transformation: Economic Strategies in Late Developing Countries (New York: Oxford University Press, 1975); Bruce Johnston, "Agricultural and Structural Transformation in Developing Countries: A Survey of Research," *Journal of Economic Literature*, vol. 8 (June 1970); and Yujiro Hayami and Vernon Ruttan, *Agricultural Development: An International Perspective* (Baltimore, Md.: John Hopkins Press, 1971) ch. 3.

2. Sholmo Reutlinger and Marcelo Selowsky, *Malnutrition and Poverty: Magnitude and Policy Options* (Baltimore: John Hopkins Press for the World Bank, 1976).

3. For a more detailed discussion of the change in relationship between developing and developed nations after the Industrial Revolution and again after World War II see Benjamin J. Cohen, *The Question of Imperialism: The Political Economy of Dominance and Dependence* (New York: Basic Books, 1975).

4. Nations in each of these categories were chosen on the basis of their impact on changing political and economic conditions in East and Southeast Asia. Intentionally avoided were many South Asian and Pacific-Oceanic nations whose food problems are similar in many ways to those in Southeast Asia, but whose analysis would require a much longer study. Australia and New Zealand also play important roles as surplus grain producers and exporters, but their food systems would also require extensive discussion.

5. The relationship of political to structural or economic conditions used here is stated quite eloquently by Michael Donnelly, who suggests that "abstract formulas do not make for automatic computation or eliminate the need for political compromise." And yet, "the process of decision is not a closed, autonomous system isolated from the economic changes which give rise to many political issues in the first place and under certain circumstances determine political outcomes." See his "Setting the Price: A Study in Political Decision Making," in T. J. Pempel, *Policymaking in Contemporary Japan* (Ithaca, N.Y.: Cornell University Press, 1977).

6. For a more detailed discussion of this strategy see Bruce Johnston, "The Japanese 'Model' of Agriculture Development: Its Relevance to Developing Nations," in Kazushi Ohkawa, Bruce Johnston, and Hiromitsu Kaneda, *Agricultural Economic Growth: Japan's Experience* (Tokyo: University of Tokyo Press, 1969).

7. There have been many studies as of late that might be mentioned here, but three good ones that criticize the Western model for its reliance on market forces are Cheryl Christensen, "World Hunger: A Structural Approach," *International Organization*, vol. 32, no. 3 (Summer 1978); Irma Adelman and Cynthia Morris, *Economic Growth and Social Equity in Developing Countries* (Stanford, Calif.: Stanford University Press, 1973); and Keith Griffin, *The Political Economy of Agrarian Change* (Cambridge, Mass.: Harvard University Press, 1974). For a critique of the socialist model see Harry Johnson, *Economic Policies toward Less Developed Countries* (Washington, D.C.: The Brookings Institution, 1967).

8. Of course, aggregate income statistics may distort actual food distribution patterns as well as the severity of food security problems. In some cases, e.g., Indonesia and the Philippines, a significant portion of the citizenry lives outside the national economy. Undeniably these people have a political impact, if not an economic impact, on policymakers.

9. For a more detailed review of conditions in individual Asian nations see Trilok N. Sindhwani, *Economic Feasibility of an Asian Common Market* (New Delhi: Sterling Publisher Private LTD, 1975), and the U.N. Economic and Social Council, *Economic and Social Survey of Asia and the Pacific, 1976* (Bangkok, 1977).

10. Sindhwani, *Economic Feasibility*.

11. U.N. *Survey of Asia*, Summary Statement, E/5980 (June 8, 1977) p. 7.

12. Johnston and Kilby, *Agriculture and Structural Transformation*, pp. 212–13.

13. Ibid., p. 213.

14. Ibid., p. 217.

15. For a more detailed discussion of Japan's domestic agriculture politics after World War II see Haruhiro Fukui, "The Japanese Farmer and Politics," in Isaiah Frank, ed., *The Japanese Economy in International Perspective* (Baltimore, Md.: John Hopkins University Press, 1975).

16. See, for example, Robert Paarlberg, "Shifting and Sharing Adjustment Burdens: The Role of the Industrialized Food Importing Nations," *International Organization*, vol. 32, no. 3 (Summer 1978).

17. For a more detailed discussion of rice price setting policies in Japan see Michael Donnelly, "Setting the Price of Rice," in Pempel, *Policymaking in Contemporary Japan*.

18. Stanley Karnow, "East Asian in 1978: The Great Transformation," *Foreign Affairs* 57, no. 3 (1979):610.

19. See for example, "Mansfield Warns Japan on Surplus in Payments Balance," *Los Angeles Times*, 5 April 1978.

20. The USDA reports that China exports a small amount of agribusiness agricultural machines like 2–5 horsepower tractors, along with rice, silk, tea, tobacco, peanuts, eggs, fruits, and vegetables. In 1976 agriculture exports comprised about 40 percent of China's exports and in 1977 rice exports earned $215 million and $350 million in 1978. China's biggest export market is Hong Kong. See USDA, *The Chinese Connection: U. S. Agricultural Trade Background on Chinese Agriculture* Issue Briefing Paper, no. 13, 24 May 1979.

21. See "More Than They Deserve," *New York Times*, 15 October 1979; and "Cambodia: A Race against Tragedy" *The Christian Science Monitor*, 17 October 1979.

22. See, for example, Wen Rong, "Food Grains for China's Millions," *China Reconstructs* (April 1980).

23. This figure was established by the author after consideration of numerous demographic estimates. The rate is generally applicable to urban and suburban areas, while "traditional motives for having several children . . . still persist in the countryside." See A. Doak Barnett, *China and the World Food System* (Washington, D.C.: Overseas Development Council, 1979), pp. 34–35.

24. Many writers, including China experts, make this claim. See, for example, Wen Rong, "Food Grains for China's Millions"; Gordon Bennett, *Huadong: The Story of a Chinese People's Commune* (Boulder, Colo.: Westview Press, 1978), p. 2; Gunter Augustini, "Agrarian Reform in China: Objectives, Approach and Achievements," in FAO, *Land Reform, Land Settlement and Cooperatives*, Rome, 1974; and the U.S. Congressional Committee on Science and Technology, "The Role of Science and Technology in China: Population/Food Balance," *Report of Subcommittee on Domestic and International Scientific Planning, Analysis and Cooperation*, House of Representatives, 95th Congress, 1st Session, 1977.

25. Nick Eberstadt, "Has China Failed?," *New York Review of Books* (5 April 1979); "Women and Education in China: How Much Progress?," *New York Review of Books* (19 April 1979); and "China: How Much Success?," *New York Review of Books* (3 May 1979).

26. For a more detailed discussion of grain output trends in China see Robert Field and James Kilpatrick, "Chinese Grain Production: An Interpretation of the Data," *The China Quarterly* 74 (June 1978):369–84.

27. Benedict Stavis, *The Politics of Agricultural Mechanization in China* (Ithaca, N.Y.: Cornell University Press, 1978).

28. Field and Kilpatrick, "Chinese Grain Production," p. 328.

29. Bruce Reynolds, "Two Models of Agricultural Development: A Context for Current Chinese Policy," *China Quarterly* 76 (December 1978):863.

30. USDA, *People's Republic of China Agricultural Situation: Review of 1976 and Outlook for 1977*, Economic Research Service, Report no. 137, p. 17.

31. From 1949 to 1952, while the Communists were still securing control over China, land ownership patterns changed slowly. As they gained greater control the Communist eliminated the "landowning elite" but preserved the "rich peasantry." According to Mao, "there should be a change in our policy towards the rich peasants, a change from the policy of requisitioning the surplus land and property of the rich peasants to one of preserving a rich peasant economy in order to further the early restoration of production in rural areas. This change of policy will also serve to isolate the landlords while protecting the small peasants and those who rent out small plots of land." Quoted in Stavis, *The Politics of Agricultural Mechanization in China*, p. 25. See also Tanaka Kyoko, "Mao and Liu in the 1947 Land Reform: Allies or Disputants?," *China Quarterly*, no. 75 (September 1978).

32. See especially chapter 3 of Stavis, *The Politics of Agricultural Mechanization in China.*

33. Alexander Eckstein, *China's Economic Revolution* (Cambridge, Mass.: Cambridge University Press, 1977).

34. Gunter Augustini, "Agrarian Reform in China," p. 31.

35. Reynolds, "Two Models of Agricultural Development."

36. Reynolds, ibid., notes that "by 1966 two-thirds of the value of agricultural machinery . . . came from the local medium and small plants." p. 868.

37. Many news services have reported changes in Chinese life-styles that have accompanied China's new modernization strategy. See, for example, "A Peking Scene: Pimps and Prostitutes at the Peace Cafe," *New York Times*, 12 October 1979; "China Seeks to Absorb Capitalist Ways Quickly," *Los Angeles Times*, 4 November 1979; and "China Opening an Ideological Door," *The Christian Science Monitor*, 13 March 1979.

38. For a more detailed discussion of China's military modernization goals see Michael D. Eiland, "Military Modernization and China's Economy" *Asian Survey*, vol. 17, no. 12 (December 1977).

39. Quoted from Seth Lipsky and Raphael Pura, "Indonesia: Testing Time for the 'New Order,' " *Foreign Affairs* 57, no. 1 (Fall 1978):199.

40. The *Beijing Review* reports that the new idea for incentives and different incomes for all types of workers is justified on the basis of "going forward" or "acknowledging the difference, opposing absolute egalitarianism and allowing and encouraging members of some advanced units with higher collective incomes to earn more . . . so as to inspire the less advanced to follow their example." See *Beijing Review*, vol. 22, no. 9 (2 March 1974).

41. Kenneth R. Walker, "Grain Self-Sufficiency in North China: 1953–1975," *China Quarterly* 71 (September 1977):565–66.

42. "Chinese Agriculture: Teams, Not Communes, Are Key," *Christian Science Monitor*, 27 March 1979, reports that agricultural surpluses are to be "shifted into local industry and public-works projects in a fashion that recompensates the agriculture sector, rather than let the peasant's incentives be damaged."

43. See, for example, "The Farmer's Market: It's Where the Action Is in China," *Los Angeles Times*, 4 November 1979, and "Canton Fair Improves Its Appeal to Buyers," *Christian Science Monitor*, 27 March 1979.

44. For a more detailed discussion see G. William Skinner, "Vegetable Supply and Marketing in Chinese Cities," *China Quarterly*, no. 76 (December 1978).

45. USDA, *People's Republic of China: Review of 1978 and Outlook for 1979*, supplement 6 to WAS 18. See also "China's New Agricultural Organizations," *The China Business Review* (September-October 1979), p. 70.

46. See "China's Harvest Hangs in Balance," *The China Business Review* (July-August 1979).

47. See "China Puts Brakes on Modernization Drive," *Los Angeles Times*, 18 March 1979.

48. Chalmers Johnson, "The New Thrust in China's Foreign Policy," *Foreign Affairs*, vol. 57, no. 1 (Fall 1978).

49. Reutlinger and Selowsky, *Malnutrition and Poverty*.

50. Lipsky and Pura, "Indonesia: Testing Time."

51. Asian Development Bank, *Rural Asia: Challenge and Opportunity* (Tokyo: The Asian Development Bank, 1978).

52. J. D. Drilon, *Southeast Asian Agribusiness* (Tokyo: Asian Productivity Organization, 1975), p. 9.

53. See William Overhold, "Land Reform in the Philippines," *Asian Survey*, vol. 16, no. 5 (May 1976).

54. Lipsky and Pura, "Indonesia: Testing Time."

55. U.N. *Economic and Social Survey of Asia and the Pacific*, 1976.

56. Indonesia, for example, an oil exporter, benefited from high oil prices and ran a payment surplus until the international economic crisis of 1974 and bad economic planning reversed its situation. Indonesia went $10 billion in debt to a number of international banks. See Lipsky and Pura, "Indonesia: Testing Time."

57. The Asian Development Bank, *Rural Asia*.

58. See, for example, Cynthia H. Enloe, "Ethnic Diversity: The Potential for Conflict," in Guy Pauker, Frank Golay, and Cynthia Enloe, *Diversity and Development in Southeast Asia* (New York: McGraw-Hill, 1977).

59. Johnston and Kilby, *Agricultural and Structural Transformation*.

60. See, for example, "Cambodia: All Sides Use Food 'Weapon,' " *Christian Science Monitor*, 10 December 1979.

61. Joachim Singlemann, *From Agriculture to Services: The Transformation of Industrial Employment* Beverly Hills, California: Sage Publications, 1978).

62. Hayami and Ruttan, *Agricultural Development*.

63. On the other hand the Asian Development Bank, *Rural Asia*, believes that some countries could benefit from a consolidation of land holdings.

64. An earlier attempt at regional organization was made by the Japanese, who occupied much of East and Southeast Asia before World War II called The Greater Asian Co-Prosperity Sphere.

65. Sindhwani, *Economic Feasibility of an Asian Common Market*, p. 131.

66. Shee Poon-Kim, "A Decade of ASEAN, 1967–1977," *Asian Survey* 17, no. 8 (August 1977):754.

67. For a more detailed discussion see Jusuf Wanandi, "Politico-Security Dimensions of Southeast Asia," *Asian Survey*, vol. 27, no. 8 (August 1977).

68. William S. Turley and Jeffrey Race, "The Third Indochina War," *Foreign Policy*, no. 38 (Spring 1980).

69. U.N. *Survey of Asia, Summary Statement*, E/5980.

70. John Wong, *The ASEAN Economies: Development Outlook for the 1980s* (Singapore: Chopmen Enterprises, 1977), p. 46.

71. See "Five Members of Asian Bloc Form Rice Stockpile for Emergencies," *New York Times*, 31 August 1979.

72. Sindhwani, *Economic Feasibility*, p. 119.

73. Shee Poon-Kim, "A Decade of ASEAN."

74. Enloe, "Ethnic Diversity."

75. "China's Good Credit Entices Foreign Bankers," *Christian Science Monitor*, 27 March 1979.

7

Agricultural Constraints and Bureaucratic Politics in the Middle East

MARVIN G. WEINBAUM

Only belatedly have agricultural development and food policy gained prominence in the developmental planning of most Middle Eastern countries. Either because schemes of economic growth were linked to oil income or because the paths to domestic prosperity were conceived in rapid industrialization, the agricultural sector rated secondary attention and, in some cases, was allowed to languish. Agriculture was not invited to make strong, competing claims on scarce national resources. By the early 1970s, probably no more than 15 percent of the public investment in the Middle East was going for agriculture. In general, policymakers expected rural areas to play a supportive role in building a modern industrial society by contributing raw materials, manpower, and food. Dams and other rural infrastructural projects deemed vital to increased agricultural production were often designed to facilitate industrialization and urban growth. Periodically, official concern was voiced about the need to raise food production and solve rural unemployment. Yet these problems were expected to ease with the introduction of advanced farm technologies, and through accelerated urban migration and agrarian reforms. New seed varieties, increased fertilizer use, and mechanization would create more intensive cultivation and free labor for the cities, while the displacement of absentee landlords and traditional tenurial arrangements would presumably strengthen incentives to increase agricultural output.

During the 1970s, most states of the Middle East were obliged to take a second, hard look at their agricultural sectors. There emerged a better appreciation of the limits of technical and social change. Without abandoning ambitions for industrial growth, many governments gave increased emphasis to rural society in overall development strategies. Industrialization had proved for most countries a far more illusive goal than had been conceived earlier. Plans were often unrealistic and projects mismanaged. Capitalization was sorely inadequate in several of the region's leading states, and long-term foreign indebtedness became identified as a serious burden. Efforts to induce rapid change had usually carried strong inflationary pressures and other strains to domestic economies. Worker productivity remained low in the region, a result of poor skills and motivation. Industrial goods were thus uncompetitive in international markets and domestic demand normally too weak to absorb most finished products. Social problems emerged out of the concentration of transplanted labor in the cities, and widening income disparities between urban and rural areas increased social and political tensions. Meanwhile, more disposable income among urban dwellers stimulated food consumption and altered eating patterns.

An expanding population, regionally between 2.5 and 3.5 percent per annnum, has served to increase pressures on domestic food stocks. Per capita food production registered a sharp decline in several countries of the region from 1966 to 1976, and for the largest number barely managed to keep ahead of population growth (see Table 7.1). Much of the optimism born of the Green Revolution of the 1960s has waned. The relatively easy gains from improved seeds-cum-fertilizer and water application were quickly realized in most countries. The early 1970s saw a slowing of improvements in wheat yields regionwide and, for Algeria, Iraq, Jordan, and Morocco, a notable drop-off (see Table 7.2). Land redistribution has frequently failed to improve the lot of small farmers or to induce more modern cultivation practices. Future increases in agricultural production appear likely to be incremental, possible only after more realistic planning and heavier public investment.

All the same, agricultural exports remain a major source of foreign exchange for many of the region's non-oil-producing states. Despite a declining share of the Gross Domestic Product, the agricultural sector continues to make a substantial contribution to national wealth in Afghanistan, Sudan, Egypt, Morocco, Pakistan, Tunisia, and Yemen PDR, among others (see Table 7.3). Almost everywhere, as Table 7.3 indicates, a sizable proportion of the labor force remains tied to agricultural pursuits. Only in Israel and Lebanon is the share of labor in agriculture below 20 percent; and in more than half the countries examined it stands higher than 50 percent.

Natural Potentials and Constraints

Attempts to understand the challenges to successful agricultural development and food policies in the Middle East usually begin with the fact that nearly everywhere in the region arable land is severely limited and water insufficient. To be sure, the Levantine coast, Egypt's Nile Valley and Delta, the Caspian Sea and Black Sea regions of Iran and Turkey, and Pakistan's Indus River Basin are notable exceptions. Other areas, especially the Tigris-Euphrates system and the Sudan's Nile savanna below Khartoum and the semitropic south, can support major soil restoration programs and

Table 7.1   Food Production Per Capita Indices for Middle Eastern Countries
            (1961-1965 = 100)

Country	1966	1976
Afghanistan	97	96
Algeria	64	87
Egypt	101	105
Iran	108	121
Iraq	100	90
Israel	99	138
Jordan	86	38
Lebanon	107	102
Libya	121	117
Morocco	93	108
Pakistan	99	117
Saudi Arabia	99	102
Sudan	94	117
Syria	73	104
Turkey	104	119
Yemen AR	91	104
Yemen PDR	99	99

Source: FAO Production Yearbook, 1976, Vol. 30.

transformation into regions of extensive cultivation. But overall, the climate and soils of the Middle East are not hospitable to agriculture intended to support large populations.

By some calculations, as much as 90 percent of the Arab Middle East should be classified as desert. A far more sanguine estimate finds 1 billion of the Arab Middle East's 3.5 billion acres to have some economic potential. Still, of this acreage, just 350 million are suitable for crops, with most of the rest good only for pasturage.[1] In fact, no more than 126 million acres of Arab lands are presently being cultivated, and of this amount less than a fifth is under irrigation; most of the remainder is at the mercy of uncertain rainfall. Even land currently in production is threatened by desertification resulting from droughts and poor land usage. Additionally, problems of soil salinity that reduce yields and eventually force land out of production plague much of the region.

As Table 7.2 point out, countries of the Middle East vary considerably in their arable land areas and crop yields. The full measure of their natural potential and real limits require, however, more extended comment. The Sudan and Pakistan are usually singled out as the best candidates to become

Table 7.2   Wheat Yields and Arable Land in Middle Eastern Countries

Country	Wheat yields (kg/ha[a])			Arable land (1,000 ha)
	1969-71	1975	1977	1976
Afghanistan	978	1213	1173	7980
Algeria	614	450	500	6500
Egypt	2741	3472	3714	2690
Iran	735	929	1033	15330
Iraq	888	601	535	5100
Israel	1442	2186	2130	342
Jordan	822	422	398	1175
Lebanon	841	1053	1200	240
Libya	240	373	609	2400
Morocco	932	931	667	7400
Pakistan	1110	1320	1430	19250
Saudi Arabia	2047	2126	1800	1040
Sudan	1136	1113	1192	7450
Syria	597	916	797	5260
Tunisia	537	877	604	3250
Turkey	1308	1593	1797	24858
Yemen AR	884	1120	1000	1520
Yemen PDR	1483	1813	1908	152

[a] 1 hectare = 2.47 acres.
Source: FAO Production Yearbooks, 1976, 1977, Vols. 30, 31.

major suppliers for the region. Both have considerable arable land not yet intensively cultivated, and both possess adequate water resources. Of the Sudan's 80 million acres suitable for crop cultivation (with another 120 million for pasturage), only 17 million are currently in production and only 4 million are irrigated. The Sudan's wheat and rice output, for examples, could easily rise as much as 150 and 1100 per cent respectively by the early 1980s.[2] Pakistan, which already places 50 million acres under cultivation and two-thirds of these irrigated, normally produces sizable rice and cotton surpluses and had been expected to be self-sufficient in wheat by the late 1970s.

In fact, neither country is likely to approach its estimated potential in the foreseeable future. In order to bring new lands into production or improve acreage in use, both countries are heavily dependent on outside assistance for development funds. Saudi Arabia and Kuwait have promised generous

Table 7.3   Gross Domestic Product and Labor Force in Agriculture in
            Middle Eastern Countries

Country	Percent of GDP in agriculture		Percent of labor force in agriculture	
	1960	1976	1960	1977
Afghanistan	—	55	85	—
Algeria	21	7	67	53
Egypt	30	29	58	52
Iran	29	9	54	41
Iraq	17	8	53	42
Israel	11	8	14	8
Jordan	16	14	44	28
Lebanon	12	—	38	12
Libya	14	3	53	20
Morocco	29	21	91	53
Pakistan	46	33	76	55
Saudi Arabia	—	1	72	62
Sudan	58	41	86	79
Syria	25	17	54	49
Tunisia	24	21	57	43
Yemen AR	—	—	83	76
Yemen PDR	—	23	70	61

Source: World Bank, World Development Report, 1978, 1979. FAO Production
Yearbooks, 1972, 1977, Vols. 26 and 31.

support to the Sudan, as have 11 other Middle East nations. But by the late
1970s regional aid to the Sudan had slowed and significant technological and
labor problems had been encountered. Pakistan's handicaps include serious
water management and administrative problems, as well as formidable social
and political impediments.[3] At present Pakistan's farmers obtain among the
lowest crop yields in the world. With a population of 80 million and a
birthrate of more than 3 percent per annum, Pakistan has had also to cope
with a sharply rising demand. Thus Sudan and Pakistan remained net
importers of food during the late 1970s. The Sudan was forced to import a
quarter of its needs and, during 1977–78, Pakistan had purchased abroad or
received on a loan basis over 1.1 billion metric tons of wheat.

Three countries in the region—Egypt, Israel, and Lebanon—have already
realized high levels of production in valuable cash crops. Lebanon with
fruits, vegetables and tobacco, Israel with fruits, dairy and poultry, and

Egypt with long-staple cotton have managed to cultivate limited acreage intensively. Yet none offers a strong potential for increasing output since their water and land limits have been largely reached. New lands are likely to be of only marginal quality, opened for cultivation only after expensive public investment. No breakthroughs of a technological kind are foreseen. Israel already employs the most advanced methods, but hopes for cheap desalination of sea water have not been realized. Lebanon feels the pressures of population against the land and competing claims on water resources. The effects of civil strife on agricultural exports, as discussed later, may be more than temporary. Despite the Aswan Dam, Egypt's agriculture picture has not markedly improved. Burdened by a population of over 40 million, Egypt has a man-to-farmed-land ratio that for 70 years has continued to drop. Costly land reclamation has not paid off and, since the late 1960s, production of major crops has declined or stagnated.[4] Policies of heavy dependence on agricultural exports by Egypt, Israel, and Lebanon no longer appear so viable as long-term economic strategies.

The presence of a substantial agricultural base but low productivity leaves considerable room for agricultural expansion in several countries of the region. Each is likely to approach self-sufficiency in at least one major crop, and exportable surpluses are possible in wheat, rice, cotton, sugar, and fruits and vegetables. No doubt, Iraq and Syria offer the most promise. Intelligent public investment in irrigation and drainage could greatly increase cultivable lands in both countries. The remaining nations in this group, specifically Turkey, Iran, Afghanistan, Tunisia, Jordan, and Morocco, have more limited potential. All are highly reliant on the vagaries of rain-fed agriculture, and substantial gains are possible only after many traditional agricultural practices are displaced. Quickly mounting domestic food demands, as in Iran and Turkey, could also diminish the value of any increase in production.

Aridity and poor soils leave the remaining countries of the Middle East, namely Libya, Algeria, Saudi Arabia, and the Persian Gulf emirates, with the lowest agricultural potential. Significantly, none has a major river system to allow ambitious irrigation projects. Yet because nearly all of these countries are oil-rich and, save for Algeria, have small populations, they have alternatives not available to most nations of the region. High-cost, advanced agricultural projects and supportive public works can help them meet a larger proportion of their domestic requirements and help to defray the rising costs of food imports. (Saudi Arabia's agricultural import bill reached $1 billion in 1977–78.) In Algeria, integration of agriculture with a more pampered industrial sector could better rationalize and modernize efforts in agricultural production.[5] The most prosperous Arab states have also recognized a means of assuring future sources by underwriting agricultural development in those Middle East countries with the highest potential.

While the constraints on increasing agricultural production in the Middle East are normally viewed as physical, or are described in technological, economic, or even social terms, this essay stresses that most policies are also charged by decisions that are essentially political. The choices made and the probability of success rest in large measure on government initiative, sustained commitment, and the capacity to implement decisions. Whether in setting broad economic parameters, identifying programs, or administering projects, governments in the region are conspicuously active. The kinds of decisions that officials make seem particularly bound up with efforts to enhance regime control. Bureaucratic practices and interests and the requirements of national and regional politics also figure strongly in choices. However analyzed, political constraints on policy are always present and often become paramount. Explanations that fail to deal explicitly with the political dimension are therefore likely to be inadequate.

Bureaucratic Norms and Structures

The study of agricultural and rural development policy in the Middle East cannot proceed very far without a scrutiny of public bureaucracies and their norms, structures, and strategies. In the continuing integration of agricultural sectors into national economies, governments across the region have been the prime movers in transforming market relationships and altering the incentives for agricultural production. Nearly everywhere, public actors take the lead in national efforts aimed at modernizing farm practices and modes of distribution. Plans for agrarian reform ordinarily rest with government initative or what is referred to as political will. It is hardly surprising, then, that much of the blame for the poor record of regimes in the Middle East in coping effectively with their food requirements development needs is laid to public bureaucracies. These agencies stand accused as well in the inability of regimes to realize promised rural social justice and demands for greater economic equality. Ministries and departments are familiarly criticized as administratively cumbersome and inefficient, incapable of formulating workable programs. The skills and personal interests of officials are viewed as major obstacles to the successful implementation of policy decisions. Institutional capabilities and individual motives cannot be understood fully, however, without reference to normative expectations, or apart from processes of competing interests, ideas, and ambitions.

Agricultural bureaucracies are unmistakably political focal points. As such, they offer an instructive setting in which to observe the ascendance of particular elites and the penetration of demand-bearing groups. Ministries are themselves participants in struggles for resources and power among units and levels of government. Few agricultural bureaucracies are insulated, moreover, from a nation's major national divisions and controversies. The imprints of center-periphery tensions, competing economic doctrines, and class and ethnic conflicts are sharply visible. Because agricultural ministries and departments are normally service oriented, and their programs administratively intensive, they are distinguishable from many other agencies for their mass clientele and high public exposure. Policies are thus more likely to generate suspicion, cynicism, and resistance and to figure prominently in the erosion of confidence in government authority. At the same time, regimes are increasingly judged by their performance in the agricultural sector. The principal officials of agricultural bureaucracies are as a result politically vulnerable, among the first to be pushed aside in changes of government priorities and leadership.

From appearances alone, many Middle Eastern bureaucracies are undergoing an administrative revolution. New units have proliferated, while older, restructured ones are assigned more explicit functional responsibilities, their procedures made more streamlined and defensible. Efforts are undertaken to recruit better trained personnel from a wider population, through less ascriptive means. Interagency working groups and conferences have been deliberately instituted to improve bureaucratic communications and coordination. Computer technology is increasingly available to aid policymakers, both for long-range planning and for practical problem-solving. Younger, foreign-educated personnel, now in the middle and even upper echelons, are the intended carriers of a new ethos of efficiency, honesty, and dedication.

These changes are not, of course, alone characteristic of agricultural bureaucracies. Nor are they peculiar to the Middle East, where they are found, in any event, unevenly across the region. Yet the widely observed elaboration of decision structures, the rationalization of procedures, and the officially approved norms in agricultural bureaucracies from Morocco to Pakistan so well illutrate how newer modes have, in fact, overlaid rather than displaced older attitudes and practices. Most important, adherence by administrators to more traditional patterns and the failure to harmonize these with newer ideas and institutions remain significant impediments to modernizing agricultural systems and establishing adequate food policies.

Agricultural ministries and agencies in most Middle Eastern countries continue to labor under a time-honored patrimonial system. It is a system of administrative authority in which all power emanates from a single political leader and where the influence of others is derivative in rough proportion to their perceived access to him or their share in his largesse. The system relies on informal relations where institutional roles and formal organizations matter far less than personal networks and reputations. Institutional rivalries and jealousies are endemic to the system, and their tensions are mediated, if at all, by the security offered by family and friendship cliques and by clientele relationships.[6]

In general, bureaucratic behavior is marked by defensiveness, both individual and institutional. Officials are cautious not to offend superiors or to identify themselves with decisions and programs that might fail to realize expectations. Ministries are typically engaged in a competition for funds with other units of government, not so much as means to pursue particular programs, but as an ongoing test of their standing in the bureaucratic pecking order. Officials are otherwise preoccupied with such symbols of recognition as upgrading existing posts, added new staff and cars, and securing choice residential accommodations.

Bureaucracies are wedded to a rigid seniority system that does little to enhance motivation or professional self-esteem of middle-level officials. There are seldom any systematic means of recognizing merit as the basis for promotion and other rewards, financial or nonfinancial. In the absence of agreed achievement criteria, ambiguity prevails over whether one will be judged by performance standards, by paper credentials, or by family and friendship ties. Opportunities for personal gain are expected very often to offset low salaries and poor working conditions. The elaborate, often redundant procedures designed to thwart administrative wrongdoing do create obstacles to many forms of petty corruption but are typically ill designed to prevent serious conflicts of interests. In some countries at least, allowances made for corrupt practices give government officials a deep stake in the prevailing political order and regime leaders a powerful lever against political deviation. On a more regular basis, political and personal loyalties are injected into bureaucratic personnel policies, resulting in the outright dismissal of some officials and the regular intimidation of many more.

Like other bureaucracies in the Middle East, agricultural ministries are usually subject to a policy of frequent and unpredictable transfers of administrators. Shifting people around is a continuous reminder of how those in superior positions can intervene on whim and at will.[7] The transfer of higher officials usually has direct implication for the initiation of agricultural programs. Much of the considerable time required for project identity and approval, as well as the high attrition rate of officials, results from the unavoidable personalization of policy. A strategically placed official's willingness to champion an idea or to intervene on behalf of a project is normally essential to its survival. Yet many officials hestitate to be closely associated with a policy whose failure can be used to discredit them and impede their career advancement. Conversely, a project's identification with specific individuals may increase its vulnerability. Transfers are likely to remove a program's originial defenders and replace them with officials holding different priorities, anxious to put their own stamp on policies. The frequent use of generalists known to be politically reliable, rather than agricultural specialists, in high posts increases the likelihood of divergent approaches. Some development projects are sensibly allowed to stagnate or die because, as the personal vehicles of earlier officials, they were poorly conceived. More often, programs are prematurely dropped lest someone else earn the credit for a success.

Relations are frequently strained between senior officials and their subordinates, especially ministry technocrats. The advice of middle-level officials may be sought on a regular basis and comprehensive policy recommendations submitted. But while the ideas and analyses of technocrats are often used to support or rationalize decisions, top officials rarely give credit to subordinates or, for that matter, involve them in final decisions. In any case, middle-level technocrats are certain to be ignored when their views fail to fit the policy preconceptions of ministry superiors.

Agricultural research and planning units tend to focus on technological concerns rather than broad agricultural policy issues and alternatives. Research is often unrelated to national development priorities. Much of it offers little practical advice to decision makers aside from conventional wisdom. Data-generated research may reach agencies only after a lengthy time lag, and information collection processes tend to be notoriously outmoded and poorly organized or supervised. Serviceable raw data is often unsystematized or presented in unusable form. Aside from questions of data reliability, there are also cases of deliberately exaggerated and falsified results, especially agricultural output figures, presented both to enhance the reputation of departments and to please the national leadership.

Few agricultural development proposals advance very far in less affluent countries of the region in the absence of a prospective foreign donor. Aside from the financial resources they promise and their special expertise, foreign nationals are also expected to provide coveted rewards to government officials in the form of travel and study. Foreigners can be useful, moreover, in sharing with government officials responsibility for projects. Unlike colleagues in the bureaucracy, the foreign adviser is not a claimant for credit in any successes and presents a convenient scapegoat in the event of failure. It is not uncommon for foreign experts to become involved, inadvertently, as allies in intrabureaucratic struggles. Predictably, the foreign adviser is resented by many in the administrative structure, especially those at the middle levels, who feel that their own advice is ignored. There are others who understandably fear that they will be unable to carry out a scheme should the foreign patrons withdraw. Very rarely can foreign assistance, whether offered by a national or an international agency, be forced on a reluctant government. Rather than the foreign representative encountering outright rejection, a government is likely to decline by allowing a decision to languish within the bureaucracy.

Because principal policymakers are not disposed to delegate real authority any more than their subordinates are to seek direct responsibility, important policy choices may at times be at the discretion of a single individual. In several Middle East bureaucracies, for example, a senior official has the power to determine whether a particular food commodity is to be imported and at what level. But unless motivated by strong personal interest, administrators often feel it wiser to reach no decision rather than to make one that may be regretted later. By a reluctance to act, many decisions involving food and development eventually find their way to top political levels where they come under the purview of personal advisers to prime ministers, presidents, and monarchs. There is typically little inhibition for these nonexperts to give advice; in fact, throughout the region, people in public life express competence to give their opinions on agricultural matters simply by virtue of having been members of landowning families. More generally, the elevation of decisions to the highest political levels is encouraged by leaders known to have an almost obsessive desire to demonstrate their indispensability by having all policy choices, often routine ones, appear to emanate from their office.

A patrimonial style in the Middle East normally follows a strategy of divide and rule. To ensure a high level of competition among a leader's subordinates, they are endowed with roughly equal power and given overlapping areas of authority.[8] An absence of defined responsibility fits well the system's informal modes and enhances the leader's flexibility to choose among personnel and policies.[9] Yet many administrative problems besetting agricultural policymaking are readily traceable to uncertain jurisdictional lines and competing bureaucracies. Different ministries and departments are often responsible for related aspects of the same problem. In Iran the government monopoly over the import of wheat, sugar, and dried milk was vested in one department, while another arranged for the purchase of meat and poultry from abroad, and both were outside the control of the Ministry of Agriculture, which set the guaranteed prices for domestic production of these items. For many years separate Iranian administrative structures assisted farmer cooperatives, managed public farm corporations, and dealt with agribusinesses.[10] The Egyptian ministries of Agriculture and Irrigation were both allowed to compete for control over reclaimed lands, and in Pakistan several government departments were expected to cooperate with an independent federal agency to administer water-logging and salinity control programs, without clear agreement over the division of tasks.

Predictably, rural development and food policies are regularly confounded as different bureaucratic sectors work at cross-purposes. Even at the sacrifice of overall planning and broader economic policy, senior administrators, after showing the proper fealty to the top political leadership, are allowed to pursue strategies that may vary widely in goals and philosophy. One ministry may stress the primacy of monetary inducements to larger producers while another prefers that increased public resources for agriculture should go mainly for grants, subsidies, and easier credits to small farmers. The conservative land policies of development banks frequently have little in common with agricultural development strategies advocated elsewhere in the bureaucracy. Administrative structures thus come to represent largely disparate economic interests, institutionalized at the national level, but also manifested through clientele relationships that reach into lower rungs of the bureaucracy. The conflicts and struggles that exist horizontally among national administrative units can thus be carried to the local levels. But as the lines extend from the center and become more remote from the highest political leadership, vertical administrative authority, basically informal and personalized, is strained. The kinds of motivated friendships and family networks that give bureaucracies necessary information and direction are impossible. The consequence for the execution of policy decisions is considerable.

Problems of Implementation

There is normally a wide gap between those who formulate and promulgate policies and those who implement decisions. The former often have little capacity to monitor the performance or assure compliance. The latter, in turn, are never certain of the full commitment in resources, personnel, and materials by their superiors. In either case, comprehensive government planning ordinarily bears little correspondence to its application in the fields and villages. Only a modest overlap exists between an official's supposed realm of authority and his scope of actual effectiveness. Poor communication between personnel in the capital and provinces explains much of this. The failure of intermediary organizations to transmit information up to policy-makers and down to farmers is frequently cited. A fundamental problem nearly everywhere is the shortage of skilled and motivated extension workers. But the ambivalence and lack of sustained attention of top political leaders and senior officials to their own policies is no less important. It is rare that the same energies that may go into starting up a program are also devoted to seeing it through.

Announcements of new agricultural projects or food production targets are designed to give the impression that government is both concerned and capable of inducing desired change. These activities are, however, less directives than exhortations to those who in fact can influence outcomes. It is almost as though the approval of an undertaking is synonymous with its realization. This distance between prescription and execution means that aspirations and weak gestures usually substitute for carefully designed blueprints. And what decisions are made somehow fail to filter down through the bureaucracy, least of all to the operational level. When they do they are as often as not misunderstood, ignored, or deliberately flouted.

Higher officials typically have an unrealistic set of expectations about the feasibility of policies or the prevailing conditions in the countryside. This results in large measure from the preconceptions of decision makers based on faulty data and inappropriately applied notions of development and rural life. Although officials are generally competent in economics and well versed through foreign travel and conferences in the latest technical advances, most remain unfamiliar with or insensitive to the day-to-day problems of the cultivator and the social implications of policy. In the climate of expectation and change felt during the 1970s, ministries and other government agencies were anxious to fashion new policies, too often with insufficient considera-tion of what was required for their successful implementation. Not a few agricultural schemes, including some well-publicized ventures with foreign governments, were conveniently dropped when the full dimensions of the necessary commitment was calculated.

The success of any agricultural development policy ultimately rests on cooperation, capabilities, and resources available at the district and field levels. It is here that the most formidable obstacles to the implementation of service activities, public workers, and agrarian reform are encountered. Inadequate public funding accounts in part for the observed shortcomings, but the problems are most often identified as ones of grass-roots management and organization. Agricultural extension activities, which are designed to attract small and large farmers alike to modern methods and enable them to apply farm inputs correctly, offer perhaps the most obvious illustration.

Extension services in most Middle Eastern countries are conducted by understaffed organizations that are either only minimally active or concen-trated in selected areas. For all of Iran's 60,000 villages, there were fewer than 1,000 full-time agricultural extension officers during the mid-1970s. A field extension worker in Pakistan looked after, on the average, 5,000 cultivated acres, which, even then, left about a third of the total farm area and the bulk of Pakistan's farmers without their services. Assigned personnel are, in any case, often neither well trained nor highly motivated. Few receive sufficient technical educations, and their salaries reflect their low status. Regular communications between extension organizations and research units are uncommon, making it impossible for workers to be informed of latest findings and techniques or for research groups to obtain sufficient field data. Because extension workers are frequently placed outside their own areas of residence, they are normally unfamiliar with local conditions and have often provided inappropriate solutions to farmers' problems. These workers have as a result little credibility among farmers and are often rejected as intruders by village leaders. More competent extension workers are likely to attach themselves to or be enlisted by already progressive farmers becoming, in effect, their field assistants.

Dedicated and skilled local workers are especially essential to the task of educating farmers and peasants in the social and technological aspects of rural development. On the assumption that most aspects of rural life are interrelated, governments have been increasingly attracted to the idea of bringing together all efforts in a more or less formalized rural development policy. Typically, the programs have been built around the concept of rural centers that are expected to improve the accessibility of services and supplies, local self-reliance, community leadership, and the decentralization of administration down to the village level. The goals of these integrated programs are not entirely new; indeed, some have been tried and many abandoned. But by the creation of more comprehensive development structures and an approach that calls for less "prescription from the top," it has been thought possible to overcome familiar resistance to rural change.

Many of the problems faced in implementing rural development are testified to by the experiences in Pakistan where, under Z. A. Bhutto, one of the more ambitious programs was attempted. Initially some local successes were registered, and by 1977 programs were established in 130 localities. But the blueprint created in Islamabad frequently failed to conform to the realities of the countryside; it called for a degree of community cooperation that was unrealistic and assumed the availability of competent field workers. Perhaps more important, rural development was found lacking because of weak links between the federal government's market and incentives strategies, and the provincial-directed delivery of material and technical assistance to the farmer.[11] Provincial planners found themselves hampered by federal commodity purchasing policies that failed to entice farmers to alter prevailing cultivation practices. Increased yields in the nonirrigated areas were also unlikely without complementary policies from Islamabad providing sufficient farm credit and fertilizer. Not unexpectedly, nearly everywhere that programs were established they became prey to local political pressures and the captives of provincial influentials. Still more, little was done to change the attitudes and practices of local functionaries. The programs were soon bureaucratized, subject to the same kind of administrative bottlenecks the scheme was supposed to supplant.

Public credit institutions have assumed a critical place in furthering government agricultural strategies. Without loans, small and medium farmers have little opportunity to purchase the necessary fertilizers, seeds, pesticides, and machinery necessary to farm intensively. Nearly all governments in the region have established cooperative loan societies and agricultural development banks to meet this need. Yet these typically underfinanced institutions have in many countries never fully resolved whether their principal mission is to uplift the poor farmer or to make commercially sound loans. Not suprisingly, and despite often strong egalitarian rhetoric, credit goes disproportionately to those successful farm operators who are able to demonstrate sufficient collateral and reasonable returns on their investment. Even in the few countries where farm credit is sufficient and a system exists to reach farmers with necessary financial assistance, the aid may not always be timely and it may be too restrictive. Once credit is allocated, public organizations often have little or no way to assure that the loans—usually given in kind—are in fact used for declared purposes. Thus subsidized fertilizer is not infrequently resold in the black market or is diverted to more profitable cash crops. Moreover, it may be more economical to rent out a tractor or have it used for haulage than to employ it on its owner's lands. In any case, most farmers are regularly in debt, either to public agencies or to private moneylenders who normally demand exorbitant rates of interest.

Problems of bureaucratic management and often inappropriate government economic policies are nowhere more apparent than in the operation of local cooperative societies. Most societies distribute a wide range of inputs, supervise their use, and undertake the purchase of harvested crops. They may also channel credit. Importantly, cooperatives act as a principal instrument in implementing government macroeconomic growth strategies. As the region's most highly integrative cooperatives, those in Egypt are intended to assure the government its desired quotas of export crops, surpluses for urban markets, and farmers' repayment for loans on seeds, fertilizers, and other inputs. Yet, in spite of Egypt's notable successes over time in rationalizing production and obtaining high yields, low administered prices and bureaucratic delays in delivering inputs have forfeited farmers' confidence and reduced substantially the government's ability to control crop production and marketing.

Through the years Egypt has often had many qualified professional managers. Nonetheless, the system suffers increasingly from a shortage of skilled supervisory personnel and is plagued by administrative corruption. Many cooperative officials are, for example, thought to profit from the sale of fertilizer on the black market.[12] Factionalism on cooperative society boards, often a carry-over from local rivalries, also contributes to cynicism among farmers. Yet government officials at higher levels have little interest in changing administrative patterns. Many have devoted years to building personal clientage networks within cooperative administrative structures.[13] The effectiveness of Egyptian cooperatives has also been impaired by its highly complex structure. Lines of authority are frequently unclear with the many diverse intermediary societies that lie between the Ministry of Agriculture and the 5,000 or so local cooperatives.

As in the Egyptian case, cooperatives elsewhere in the Middle East have usually followed in the wake of land distribution programs to small farmers. The scattered settlement of small, often fragmented, holdings provided little opportunity for mechanization or modern systems of seeding, irrigation, harvesting, or marketing. To overcome this, as well as to assure adequate social organization following the departure of large landlords, small farmers were expected to become members of rural cooperatives. In few countries of the region, however, has the cooperative movement acquired the scope of coverage envisioned. Despite the goal in Syria to create 4,500 cooperatives by the end of 1975, barely more than 1,700 were in place, and the rate of expansion had slowed notably in the 1970s. Iraq was able to form only slightly more than half of its planned cooperatives in the ten years after the program began.[14] In Iran, efforts to expand and redesign cooperative societies resulted in the creation of over 5,700, with more than 2.6 million members by the mid-1970s.[15] In reality, most of these organizations existed only on paper and never fully assumed the role intended. Failure in the Iranian case to gain popular approval or to contribute significantly to the growth of commercially available food supplies illustrates problems that are region-wide.

Many beneficiaries of Iran's land reform were able to avoid affiliation, and membership was sometimes, in fact, limited.[16] Officials did little to educate peasants about the purpose or workings of cooperatives, often forcing on them alien concepts.[17] Most cooperatives were too poorly funded, their credit terms too restrictive, and prices paid to farmers that were too low to encourage increased production. With hardly adequate marketing facilities and a frequently severe shortage of storage and transport, only a small fraction of the country's agricultural products moved through the rural cooperatives. But, again, the shortcomings were also administrative; chronic under-financing brought the recruitment of poorly paid, inadequately trained supervisory personnel. This deficiency was magnified when Iran's policy-makers, in an effort to find a better means to organize small farmers, inaugurated more centrally managed farm corporations—in effect, joint stock companies. Predictably, farmers balked at participation in a scheme in which they were asked to surrender recently acquired, tangible property rights for abstract shares in a state-run enterprise.

The Case of Agrarian Reform

Continuous administrative intervention and supervision is no doubt a major prerequisite for the enforcement of agrarian reform programs. In what can be, as Herring and Chaudhry observe, "a severe drain on administrative energies," provincial and local bureaucracies carry much of the burden in seeing that reform is fully and justly implemented.[18] The inefficiency and abuse often found to characterize the activities of these agencies can be used, then, to explain the failure of governments across much of the region to follow through on their designs for agrarian change. To be sure, weak political commitment from the top, as implied in faulty land redistribution and tenure reform legislation and directives, quickly registers on local officials. Many regimes, sincere about reducing rural unrest and stimulating agricultural productivity through reforms, have, in the end, backed away from direct confrontation with rural elites. Tenants and other beneficiaries of change must also take the blame for the only partial success of most programs. Farmers must assert their rights and report illegal actions if programs are to work. But their acquiescence may be entirely rational in those countries where reforms do not alter fundamental economic forces and secure their livelihood. Aggrieved farmers and the landless may also logically fail to take any initiative on their behalf when they are obliged to find redress among local officials they feel are neither impartial nor capable of protecting their rights.[19]

The administrative machinery required to enforce acreage ceilings is found especially wanting. Both inertia and unreliable data allow many landlords to avoid having to declare land in excess of legal limits. Landlords have been most successful in concealing their full holdings where these are spread over several villages. In some countries, the ceilings on ownership were kept deliberately high, and laws permitted landlords to retain large estates through various loopholes. Thus many large farms remained in private hands after Iran's 1962–1963 reforms because of their exemption as productive, modern farming enterprises. Other loopholes have allowed landowners to subdivide land among family members, friends, or individuals economically dependent on them. Misclassifications of soil, forgeries, and illegal evictions of tenants were revealed after Pakistan's 1972 reform. And, except where land has been entirely confiscated in more revolutionary programs, landlords have usually managed to retain the most productive acreage.

The opportunities for administrative corruption are numerous where land is to be distributed. Unscrupulous reform officials sometimes sell land to the highest bidder or indirectly purchase it themselves. Expropriated land is at time turned over to local individuals with good party credentials. In Iranmuch of the promising new development tracts went to individuals with sizable capital and personal access to the bureaucracy rather than to peasants. New tenancies were created in Pakistan on resumed land in order to provide additional income for provincial governments. In general, rural influentials have been able to use their influence with key functionaries to soften the blows of reform and even to turn it to fiancial advantage. These same contacts have also enable them to easily intimidate expectant small farmers and the landless.

Events in the Middle East appear to confirm that nothing short of major transformations of social and economic power in the countryside are likely to prevent landlords from frustrating reform, and only the displacement of traditional political authority, including an entrenched bureaucracy, will allow peasants to claim and retain new legal rights. Of the region's more far-reaching land redistribution efforts—those in Egypt, Sudan, Iraq, Iran, Syria, Libya, Pakistan, Algeria, South Yemen, and Afghanistan—all but those in Iran, the Sudan, and Pakistan were accompanied or preceded by major political upheavals and noteworthy shifts in the distribution of power.[20] Even so, a more privileged class of farmers has survived in almost every country. Moreover, the most recent agrarian reforms, those in Afghanistan, also indicate how fundamental changes may create new administrative problems.

In the determination of the shaky Afghan government to win over the sympathies of the rural masses to the April 1978 Revolution, the Marxist leadership initiated rural reform by waiving all debts and mortgages hanging over the heads of small farmers. The new regime followed up this largely popular move with the inauguration of a massive program to redistribute the property of large estates throughout the country. Rather than entrust the administration of this effort to provincial civil servants or create a land reform bureaucracy, the Khalqi leadership detailed more than a thousand of their party faithful to the countryside. By mid-1979 these workers, most of them young urban ideologues, had directed the distribution of land to some 180,000 families.[21] But in the haste to bring revolutionary change in the rural areas, they produced a program that quickly courted economic disaster and proved to be a dubious way to win popular loyalties. Uncertain which land would be confiscated, more-prosperous farmers refused to cultivate fully their fields or provide the accustomed credits to smaller farmers. In the absence of adequate land surveys, undemarcated seized land, perhaps as much as 15 percent of the cultivated area, went unsown.[22] Those peasants who benefited from the changes were awarded a barely economical two and a half acres of first grade land or its equivalent in larger, poorer quality parcels. Despite promised formation of cooperatives to assist these farmers, little thought was given to their organization or funding. With no institutional means to obtain seeds and fertilizer, many new land owners felt deserted and cheated by the government.

For all its mistakes, the Afghan reform effort had idealistically sought to use local committees to involve peasants in the land redistribution process. Whatever a government's ideological bent, there is across the region increasing understanding of the argument that greater popular participation in decisions is necessary to successful rural development as well as reform. Opportunities for peasants to gain an awareness and to participate in decisions affecting them are not infrequently built into rural programs. Local involvement institutionally takes the form of councils where farmers or their representatives are supposed to be able to influence the shape of public projects. By giving farmers some say in the allocation of resources at the local level and decentralizing the responsibility for determining priority activities, the locals are expected to feel greater incentive to cooperate and be less suspicious of new ideas. Participation should also help in making more informed choices at higher government levels.

Decentralization, all the same, runs counter to the historical trend in most countries, in which central governments have struggled against local and regional elements to create more economically and politically integrated systems. Many planners fear that local communities, delegated greater authority, will be better able to resist innovations, and argue that those at the local level can be every bit as corrupt, wasteful, and misinformed. The idea of sharing initiative and responsibility with farmers and other rural dwellers in fact meets strong opposition throughout government bureaucracies. In practice, administrators almost everywhere continue to monopolize the management of agricultural development and reform activities. Officials view almost any increase in the role of popular bodies as a net loss of their powers and a diminishing of their status. Efforts to widen participation and control are also vitiated by the many people, in and outside of government, whose vested interests depend on the concentration of policymaking discretion in the capital. At most, then, farmers are allowed to engage in a form of pseudoparticipation. Local councils soon lose their credibility among peasants, as they are found to be essentially a means for government functionaries to legitimize decisions reached elsewhere or are used primarily to monitor local criticism of the regime.

Policy Processes and Styles

If a public policy consists of a defined body of coherent objectives and a consciously derived set of strategies, then few governments in the Middle East can be said to have a policy for food and agricultural development. In place of comprehensive policy analyses that identify targets and create a set of projects designed to realize these goals, projects are more likely to be assembled ad hoc and specific budgetary allocations arrived at without reference to the consequences of various plans or their consistency with national economic objectives. Few bureaucracies in the region, moreover, have the means or inclination to monitor and evaluate the impact of programs in a way that would aid top policymakers to make adjustments. Indeed, the very design of programs often makes it difficult to assess their progress.

In a manner not peculiar to agriculture or the Middle East, there is considerable comfort taken in forming new administrative units in the belief that, somehow, by the creation of an agency, a problem has been addressed and is on its way toward being solved.[23] Some units are thus disbanded or bypassed while newer ones are formed with almost identical goals and with no more authority or commitment of resources than the structures they replaced. The purpose is as often as not to break up concentrations of power that threaten to upset the balance of rivalry and conflict that reinforces the leaders' sense of control and policy dominance. At other times, however, the erection of units formed outside the traditional bureaucracy is a conscious decision by leaders to overcome known rigidities and to advance more quickly better qualified people. Rural development and new lands agencies have in several countries been established to circumvent old-line ministries and to group logically related activities. But unless these newer agencies continue to attract the special patronage of high political authority, their funds will in time be diverted and their powers sapped by traditional departments. In any case, once bureaucratic relations become firmly established in the newer units, they are likely to assume most of the same patterns of patrimonial behavior as is found elsewhere in the bureaucracy.

One of the more notable features of policy style among top planners and political leaders in the region is an impatience and even an intolerance for incremental change. For reasons perhaps as much psychological as immediately political, the preference is for quick, dramatic results. In this continuous search for panaceas, projects are likely to be dropped prematurely in favor of new, often imported, solutions. Program failures seldom serve as learning experiences in which workable ideas are retained and others disgarded. A good argument can be made that this tendency toward abrupt changes in policy is increased by a lack of meaningful participation by affected publics.[24] In the absence of public accountability that would bring a continuous questioning of decisions and require a grass-roots testing of results, officials are unrestricted from sudden new departures in approach and emphasis. The disinclination to improve rather than displace programs is illustrated on a grand scale in Egypt's Open Door policies begun in the mid-1970s. Instead of dealing with shortcoming of existing policies, including the strengthening of institutions long in place, the Sadat government turned its back on much of the Nasser socialist legacy. In the hopes of gaining an illusive economic miracle, policymakers sought to introduce a set of market-oriented investment and production priorities, encouraging private, profit-maximizing firms to compete with the publicly managed agricultural sector.[25]

Ideology has seldom served as a strict guide in government decisions involving agriculture. If it had, no doubt a more consistent set of development policies would have emerged in most countries. A frankly capitalist agricultural strategy would provide for a maximization of incentives for producers by eliminating government interventions in the form of price controls, noncompetitive procurements, export constraints, import subsidizations, and restrictions of farm size and land tenure. Through the lure of higher prices for crops, the most enterprising farmers would be stimulated to increase their output. These policies would be expected to encourage modernizing rural elites to increase private investment, and in the process to buy out smaller producers, introduce modern management, fully mechanize and, in general, take advantage of economies of scale. The "trickle down" mechanism that is eventually supposed to increase the national well-being would unavoidably lead to wider disparities in income. It would, however, reduce urban-rural sector differences, and the expected expansion of production should employ more people in the countryside, even with increased mechanization. The consumer, though paying more for food in the short run would, overtime, be an expected beneficiary of more plentiful domestic production and higher national growth rates.

Alternatively, an uncompromising socialist strategy would allow state control of planning and agricultural investment, and farm output to be bought and sold by the state at fixed prices. The system would stress the removal of economic privileges through radical agrarian reforms that destroys the traditional class structure. Along with tenancy and sharecropping, there would be no place for the small independent farm, which would be phased out in favor of government-managed cooperatives or state farms. The strategy would aim for a more equitable distribution of rewards, and would expect to attain higher levels of agricultural output through better rationalization of the modes of production, the introduction of new technologies, and the release of previously untapped human resources.

While both systems have their advocates in the region, no regime has consistently followed one set of agricultural development strategies or the other. Rather than a socialist or capitalist model, ruling elites have almost everywhere adopted policies that contain elements of both. Thus countries with large private sectors regularly intervene to control selected crop prices and consumer food costs, and countries with bulging public sectors decline to monopolize production or relations of exchange. The national leaders in the Middle East and elsewhere are apt to insist that theirs is a "third" or "middle" way. As described by John Osgood Field, the "middle way" "rejects the evolutionary progress of Western societies at the same time as it seeks a viable alternative to the disruptive, revolutionary prescriptions of the Communist model. It seeks an immediate, focused response to a particular program without waiting for development to 'trickle down' and without requiring a wholesale transformation of the nature of society."[26]

Historical experiences with Western imperialism as well as the continuing popular aversion to anti-Islamic doctrines associated with the Soviet system naturally leave regimes reluctant to try to impose unmodified foreign blueprints. Policies that reject the incorporation per se of economic theories and practices of the West or Communist countries ideally should allow countries of the Middle East to avoid unnecessary doctrinal binds and enhance their chances of policy flexibility and adjustments suited to national differences. Yet, as described above, bureaucratic decision styles militate against an incremental approach that experiments with various schemes inspired by the two models. In the willingness of policymakers to accept the utility of aspects of both the capitalist and socialist systems, they have created not so much a synthesis as they have often impetuously picked piecemeal or alternated among largely incompatible approaches.

The anomalies in agricultural and development policy are plainly visible, helping to account, at least in part, for half-hearted, self-defeating national efforts. These outcomes are a predictable consequence of trying to reserve a critical role, in some countries, for the private sector while following an interventionist path that encourages severe distortions of a free market. Despite the biases of some governments toward more successful, private agricultural enterprise, their policy instruments are often blunted by uncoordinated, uncertain actions. Even among regimes concerned with the income of smaller producers, strategies designed to provide the strong incentives necessary to stimulate output break down in ineffectual delivery of farm inputs or in deference to urban consumerism. Guaranteed prices that are supposed to make a farmer's livelihood more secure only manage to stifle what initiative exists. A new class of private holders may be created in countries committed to agrarian reform only to allow them to succumb economically because governments fail to allocate the necessary resources or prove unwilling to incure the political costs of enforcing participation in cooperative goals. For governments that otherwise are committed to managed economies, there is surprisingly little consideration for behavioral effects on farmers of the price relationships among crops or price differences between domestic and international markets. In the failure to find coherent mixed strategies, these regimes often ignore the logic of both the capitalist and socialist models and deny themselves the presumed advantages of each.

If not a coherent agricultural policy, every bureaucracy in the Middle East can be said to affirm a regime's distinct interest biases. Traditional landed families retain their economic and political preeminence through administrative patronage in present-day Jordan and Pakistan, while in Egypt a stratum of wealthy peasants occupies a privileged position in public allocations and access to government functionaries. During the Shah's reign, private large-scale farming was promoted by key national ministries, and resources have gone disproportionately to bureaucratically organized small farmers in Algeria and Libya. The administrative apparatus need not always be an active or consistent instrument of classes or groups in order to serve as guarantor that their values and goals will prevail. Ruling elites are sometimes obligated to advocate policies that promise largely universal class and group benefits, and on single issues a regime may be in conflict with usually allied interests. It is difficult to obscure, however, an overall partiality in government priorities and bureaucratic interventions or, most fundamentally, the state's protection of an economic and social order to which favored interests owe their survival.

Conclusion

Even were the hurdles to more rational and effective bureaucratic policy-making cleared, the reality of mounting food deficits in the 1980s would remain for the region. Limited and unrealized production assures that few if any countries in the Middle East are destined to become self-reliant in meeting their domestic food requirements. Earnest national efforts may be able to better exploit available land and water resources and to provide more adequate economic incentives for cultivators. New modes of organizing farmers can help optimize agricultural potentials, while dedicated agrarian reform if linked to comprehensive rural development, promises to increase productivity. Even granting such changes, reliable sources of foreign food trade and assistance will continue to be indispensable across the region. Without Western creditors, the less affluent countries face certain food shortages and probable political unrest. States with ample oil revenues may feel more confident in their ability to purchase needed farm commodities, equipment, and expertise in international markets. But rich and poor alike are uneasy with the extent of their dependence on sources outside the Middle East. Neither foreign trade nor foreign aid is viewed as entirely reliable, and both frequently carry implicit political expectations along with explicit economic preconditions.

Prospects for reducing extraregional dependencies rest in large measure on increasing economic integration and cooperation among Middle Eastern states. The oil-exporting countries of the region have the accumulated capital to assume a larger share of the financial assistance and development aid required by their less affluent neighbors. Intraregional transfers of skilled and unskilled labor and valued remittances to home countries are already dominant features of the Middle East and will continue to underpin development plans and relieve budget problems. But the centerpiece in a complementary food policy for the region would be a system of free trade where countries of the region specialized in cultivating those crops that they produced most efficiently. Collaboration would also logically extend to the manufacture of farm machinery, fertilizers, and other inputs, as well as a sharing of agricultural research.

If the recent past is any guide, regional integration of economies is at best only a distant goal. At present agricultural trade among countries of the Middle East is modest; the flow of imports and exports is, in the main, with the Western countries. Of the five Arab states that were once expected to constitute a Common Market, Egypt, Iraq, Syria, Jordan, and Kuwait, only Jordan found its major trading partners among Market members; for the rest, after all, much that is uncomplementary about the agricultural systems of the region, and the job of reorienting trade would be formidable. Not only do they presently compete in many food crops, but countries of the region often have highly contrasting agrarian systems, ranging from largely unfettered private enterprise to state-dominated rural economies.

Regional cooperation seems improbable while countries of the Middle East remain engaged in bitter national rivalries and affected by deep mutual suspicions. The few serious attempts to coordinate economies have occurred in the context of pan-Arab inspired schemes of political unification. These comprehensive plans naturally collapse when political marriages either fail to be consummated or are short-lived. Better prospects for cooperation would seem to lie in economic agreements of a limited variety, between countries acting primarily on the basis of economic convenience rather than political motivation. The impetus for forging common, practical economic bonds may have increased with the widening concern in the Middle East that food trade and development aid have already become important levers in international politics. The continuing stagnation of domestic agricultural production, in no small part traceable to almost immutable bureaucratic constraints, should also prompt a search for regional solutions.

Notes

1. From a study by Egyptian agronomist Mustafa al-Gabali, cited in the *Christian Science Monitor*, 27 December 1976.

2. Reported figures at the World Food Conference in Rome, 1974. See *Events*, 13–19 1976 pp. 38–41.

3. An extensive treatment is contained in M. G. Weinbaum, "Agricultural Development and Bureaucratic Politics in Pakistan," *Journal of South Asian and Middle Eastern Studies* 2, no. 3 (Winter 1978): 42–62.

4. John Waterbury, "Aish: Egypt's Growing Food Crisis," Northeast Africa Series, *American Universities Field Staff Reports*, vol. 19 no. 3 (December 1974).

5. John Waterbury, "Land, Man and Development in Algeria, Part I," Northeast Africa Series, *American Universities Field Staff Reports* 17, no. 1 (March 1973): 6–14.

6. For a more extended treatment of food and agricultural development policies and the politics of the Middle East see M. G. Weinbaum, *Food, Development and Politics in the Middle East*, Westview Press, forthcoming.

7. An Extensive and valuable discussion of patrimonialism in the Middle East is found in James A. Bill and Carl Leiden, *The Middle East: Politics and Power* (Boston: Little, Brown and Co., 1979), pp. 150–177.

8. Ibid, p. 167.

9. See Robert Springborg, "New Patterns of Agrarian Reform in the Middle East and North Africa," in *The Middle East Journal* 31, no. 4 (Spring 1977): 140.

10. For a full discussion see M. G. Weinbaum, "Agricultural Policy and Development Politics in Iran," *The Middle East Journal* 31, no. 4 (Autumn 1977): 434-50.

11. See Norman K. Nicholson and Dilawar Ali Khan, *Basic Democracies and Rural Development in Pakistan* (Ithaca, N.Y.: Center for International Studies, Cornell University, 1974), p. 38.

12. A highly revealing examination of Egypt's cooperative system is found in John Waterbury, "Egyptian Agriculture Adrift," *American University Field Staff Reports*, Africa Series, no. 47 (January 1978), pp. 6–8.

13. Springborg, "New Patterns of Agrarian Reform," p. 141.

14. Hossein Askari and John T. Cummings, *Middle East Ecomonies in the 1970's (New York: Praeger Publishers, 1976), pp. 153, 159.

15. M. G. Weinbaum, *"Agricultural Policy and Development Politics in Iran,"* p. 438.

16. Jane Clark Carey and Andrew Galbraith Carey, "Iranian Agricultural and Its Development 1972–73," *International Journal of Middle East Studies* 7 (1976): 365–66.

17. Daniel Craig, "The Impact of Land Reform on an Iranian Village," *The Middle East Journal* 32, no. 2 (Spring 1978): 48–49.

18. Ronald J. Herring and M. G. Chaudhry, "The 1972 Land Reform in Pakistan and Its Economic Implications," *The Pakistan Development Review Quarterly* 13, no. 3 (Autumn 1974): 272–73.

19. These points are argued in a paper by Ronald J. Herring dealing with tenure reform in Pakistan and the countries of South Asia, "The Rationality of Tenant Quiescence in Tenure Reform: Production Relations and the Liberal State," delivered at the Annual Meetings of the American Political Science Association, Washington, D.C., 1 September 1979.

20. For descriptions of developments in most of these countries see Askari and Cummings, *Middle East Economies in the 1970's*, pp. 119–77.

21. *The Washington Post*, 11 June 1979.

22. *Middle East Economic Digest*, 4 May 1979. p. 3.

23. Richard M. Fraenkel, "Introduction," *The Role of U.S. Agriculture in Foreign Policy*, Richard M. Fraenkel, Don F. Hadwiger, and William P. Browne, eds. (New York: Praeger Publishers, 1979), p. 13.

24. See *Pakistan Economist*, 13 May 1978, p. 11.

25. For an extended treatment of the Open Door policy see John Waterbury, "The Opening, Part I: Egypt's Economic New Look," *American Universities Fieldstaff Reports*, Northeast Africa Series, vol. 20, no. 2.

26. John Osgood Field, "Needed: Nutritional Planning and Development," in Ross B. Talbot, ed., *The World Food Problem and U.S. Food Politics and Policies: 1977* (Ames, Iowa: Iowa State University Press, 1978), p. 33.

8

Tropical Africa: Food or Famine?

LYNN SCARLETT

Introduction

The complex mosaic of microclimates, cultural settings, and political environments in tropical Africa have often discouraged attempts to analyze its food situation. Yet the geographical area between the Sahara and South Africa displays enough homogeneity in food production and consumption patterns to justify analysis as a unique regional entity.

For virtually all of tropical Africa the rural sector is the largest employer of labor, and exports are comprised primarily of agricultural products. Diets consist overwhelmingly of starchy staples, and most production of food staples for domestic use is derived from small farmers operating on a partially self-sufficient basis. Patterns of land ownership vary, but a legacy of individual cultivation of plots without payment of rent often exists within a formal communal ownership structure. Finally, rapid population growth rates have universally fueled the demand for foodstuffs in tropical Africa.

Lest one become too sanguine about the uniformity of food problems facing tropical Africa a word of caution is required. With all its similarities, the area is nonetheless comprised of over thirty countries strikingly different in size and population. An astounding diversity of microclimates interrupts the otherwise predominant climatic patterns. Though political instabilities convulse much of tropical Africa, such problems are by no means universal. Nor do the various political regimes exhibit identical philosophies. Finally, in any regional or even national study one risks overemphasizing aggregate statistics and patterns, thereby disguising problems particular to individuals or isolated groups of people. However, the proposed focus on tropical African

food problems will attempt to highlight common problems without oversim-
plifying them.

Africa's Food Problem

Recent striking increases in the importation of staple foodstuffs throughout
tropical Africa would seem to give credence to a pessimistic interpretation of
Africa's agricultural production potential. But growing import figures are
not necessarily an indication of an inherent physical or cultural inability of
Africa to meet its regional food requirements. Growing imports of food
staples result more from domestic policies and political instabilities that have
impeded the production and distribution of local foodstuffs. Minimum wage
policies and rural land taxes have aggravated rural-to-urban migration
trends, increasing the demand for food in urban markets. Coupled with food
price ceilings, licensing restrictions, and rigid production quality controls of
domestic foodstuffs, the net effect has been to discourage agricultural
production altogether, or to prevent it from reaching the growing urban
markets, leaving governments to rely increasingly upon imports to meet the
demand for food.

Though agricultural production has fallen short of the theoretical potential
in tropical Africa, the food problem is not as dismal as recent claims often
insist. A look at FAO agricultural figures reveals that food output has, if
anything, increased rather than declined over the past decade.[1] It is true that
for some staple crops marketed production has increased only moderately,
scarcely keeping up with population pressures. Nonetheless, it would be
premature to point to these figures as evidence of the inherent inability of
African farmers to meet most domestic demand requirements. Such claims
serve to turn policy focus to the global forum where food aid, or improvement in the terms of trade for exports, might be forthcoming, while leading
to neglect of the internal and regional potentials for producing foodstuffs
cheaply and competitively. A number of studies staunchly support the view
that tropical Africa possesses vast potential for increasing its food production.[2] Why increased production has not materialized, or has occurred only
sluggishly in the face of this potential, is at the heart of understanding
Africa's food problem.

It will be argued that, where government policies have single-mindedly
stressed increased industrialization and reflected an urban bias, and where
agricultural markets have been manipulated and controlled or altogether
disrupted by political upheavals, agricultural production has floundered. If a
food problem exists in tropical Africa it is not primarily due to resource
constraints, cultural backwardness, or inefficient farming techniques.
Rather, ill-conceived policies, a zeal for planning where information is
lacking, and a general bias against agricultural commodity markets have
exacerbated, if not generated, the food problem. Eliminating the panoply of
government policies currently restricting the operation of both domestic and
export markets in agricultural products is thus the first step toward alleviating tropical Africa's food problem.

Most food continues to be produced and consumed within national borders. Hence, if the food problem is to be alleviated the first task must be to improve domestic production and distribution patterns by allowing markets to operate more freely. This does not suggest that self-sufficiency in food production is either the immediate or the ultimate goal. To aspire to self-sufficiency would serve merely to utilize scarce resources for the production of goods that might be bought more cheaply elsewhere while failing to exploit opportunities for the production of goods where a comparative advantage is present. However, much of tropical Africa manifests considerable potential to produce food competitively for internal consumption or in surplus to trade for other food products. Increasing domestic food production does not preclude local and regional specialization to capture the benefits of economies of scale and comparative advantages. It is where government policies have stifled the initiative to exploit such opportunities that the source for much of Africa's food problem lies.

Before embarking on the proposed analysis two caveats must be stressed. First, though national policies are much to be blamed, the international economic system should not be ignored even if less than 10 percent of staple foods consumed move in international trade.[3] A second caution deserves mention. Appeals to redirect national agricultural policies imply that such changes are politically feasible. Without appearing overly pessimistic, there is some doubt whether changes, especially those formulated to encourage agricultural market systems, will be forthcoming. This is particularly true as the proposed argument for encouraging more competitive markets does not disguise a push for the neocapitalist version of laissez-faire. The latter has often been put forward as an apology for government intervention insofar as it protected the interests of the wealthy, engendering such suspect policies as corporate tax "breaks," monopolistic licensing practices, antiunion laws, credit controls, and so on. Such policies, promoted in the name of free markets, have been as conducive to stifling the initiative of individual small farmers and traders as have been policies designed to put economic control more squarely on the shoulders of governments.

Advocating the encouragement of more freely operating markets in tropical Africa is essentially a plea for fundamental structural changes, including land reforms and the elimination of existing taxation policies and credit controls. Such structural changes would have the ultimate effect of redistributing access to income-producing assets, thereby more widely dispersing the benefits accrued from more freely operating markets.

Dimensions of the Food Crisis in Tropical Africa

In its most basic sense, "the food crisis" refers to the inability of food supplies to satisfy demand. While this general problem is self-evident, its underlying dimensions are not. Is the problem primarily the result of an absolute lack of food supplies in relation to demand? Is the problem more one of high food costs in relationship to incomes available to the hungry for expending on food? Or, even if adequate supplies of some food staples exist at affordable costs, are the "right" foods available to maintain nutritional diets? To pinpoint possible remedies for tropical Africa's food situation one must assess which of these conditions, or combination thereof, constitutes the essence of its food crisis.

Some suggest the problem lies primarily in the first consideration: Africa's population is exploding uncontrollably while land and resources available for food production are shrinking. Though this "spaceship earth" theme may pose some critical questions, the evidence to support it seems, at best, arguable.[4]

Certainly Africa's food problem has been complicated by an unprecedented upsurge in the growth in demand for food. This demand has resulted from a rapidly growing population, increasing incomes among some groups of people, urbanization, and the move from self-sufficiency to specialization. In 1969 the total African population did not exceed 360 million. Just ten years later that population has expanded to approximately 436 million.[5] This represents, for most of tropical Africa, a population growth rate averaging 2.7 percent per year, contrasting markedly from the prewar growth rate of around 1 percent, or the developed-country rate of considerably less than 2 percent.[6]

Added to the absolute increase in demand for food generated by such population growth rates are the effects on demand for marketed agricultural products engendered by the explosive growth of urban populations, estimated at 4.9 percent per year in Africa.[7] These trends complicate the food problem by requiring not only increased production but improved marketing of existing production.

But the scope of the food problem in tropical Africa cannot be fully appreciated merely by pointing to unprecedented rates of population growth. Availability of nutritional foods, food costs, income levels, and access to food-producing assets must also be considered.

It is claimed that the lack of protein, vitamins, and minerals provides cause for alarm in Africa. To assess this claim is particularly difficult because data are scarce and often unreliable.[8] However, a particularly revealing assessment of the tropical African diet concludes that "diets are those of poor people but they are not necessarily poor diets."[9]

Controversy over the regularity of a good diet presents yet another potential problem in African food consumption. It is repeatedly asserted that seasonal fluctuations in food availability impose serious shortages in the African diet. Two factors fuel this claim. First, relatively little agricultural production is traded in domestic markets, making reliance on local supplies imperative. Should shortages occur in a given locale, lack of trade prevents such fluctuations in supply from being offset by surpluses from other areas. Second, because most agricultural production centers around one or only a few different staples, it is asserted that a "hunger gap" arises between the depletion of supplies from one harvest and the availability of food from the new harvest.

The prevalence of this hunger gap has been overstated. Assessment of the purported hunger gap in Africa reveals that its persistent and widespread occurrence is doubtful. Most Africans engage in staggered plantings and mixed cropping, both of which are practices that extend the harvest period over a relatively long period of time. In addition, African peasants store produce where feasible. Perhaps most important is the fact that root crops, in particular cassava, can be harvested essentially year-round, and serve as adequate substitutes when other staples are in short supply. Finally, a range of wild plants are virtually always available for consumption in rural areas should other staple supplies prove insufficient.

Though the hunger gap does not appear to plague tropical Africa regularly, the problem is not altogether an illusory one for some areas. Especially in isolated regions that rely on a single staple crop, the potential of such a gap's occurring can be pronounced. However, such one-crop areas are relatively rare. By far the greater impetus toward creating a hunger gap has been the appalling disruptions in production and trade accompanying recent wars and political disturbances. Faced with interruptions in transportation, domestic food markets have suffered considerably, thus increasing the potential for periods of famine. To discount the prevalence of any systematic and widespread hunger gap does not, of course, preclude calamitous famines resulting from natural disasters such as the Sahelian drought, which commenced in 1968. Such famines pose severe problems for human survival and hence demand attention from both national and international relief mechanisms. However, their occurrence must be distinguished from the more general and persistent food problem currently under discussion.

The above comments suggest that tropical Africa's food problem is not primarily one of absolute deficiencies in food supplies in relation to the population. Nor does a deficient composition of diets generally afflict tropical Africa. A third view contends that the problem is essentially one of maldistribution both of income and of available food supplies between and within rich and poor nations. This view argues that, while absolute food supplies might be potentially sufficient to feed the growing African population, food distribution is imbalanced.

Certainly the problem of poverty and maldistribution of income, both on a worldwide basis and within individual African nations, is inextricably linked to the food problem. Money buys food. Moreover, increased income is closely linked to declining population growth rates.[10] There are, nonetheless, both practical and theoretical problems with focusing mainly on redistributive schemes to alleviate tropical Africa's food problem. From a practical standpoint one must ask whether purely redistributive schemes are likely given current economic and political power configurations. Even if the national political will exists to improve the distribution of incomes and food, control over the international food distribution system would be unlikely. Insofar as food imports would be required to supplement domestic food supplies in a redistribution scheme this problem becomes highly significant.

It is conceivable that the above practical problem could be overcome given the necessary political pressures. However, several fundamental theoretical problems suggest that a focus solely on redistribution is neither an effective nor a desirable means of alleviating tropical Africa's food problem. In the first place, the emphasis on redistributive policies arbitrarily separates production and consumption activities, presuming that increases in production don't necessarily improve the consumption of those who most need it.[11] The argument rests on several assumptions. First, it is maintained that in a market economy unequal distribution of assets necessarily permits only the wealthy to take advantage of production opportunities. Hence, increases in food production are likely to benefit only relatively wealthy farmers while failing to augment either incomes or food consumption among the poor. Second, it is argued that increases in production in a market economy where wealth and incomes are maldistributed take the form of more luxury items for the rich, leaving production of basic food staples essentially unchanged.

While it is doubtful whether the above assumptions are even generally valid, they are particularly misleading with regard to tropical Africa. The rural sector includes 80 percent of tropical Africa's population. It is precisely this rural community that comprises the poorest sector. In addition, income distribution within the rural sector varies considerably less than between the rural and urban sectors. There is, thus, a direct connection between improving food-staple production and increasing the incomes, and thus the food consumption, of the poor. Indeed, if the incomes of the rural poor are to improve at all, increased agricultural production to provide surplus for sale is essential. Increased agricultural production is intimately connected to improving rural incomes and alleviating tropical Africa's food problem. The redistributive approach fails to appreciate that connection, especially insofar as the tropical African poor farmer is concerned. Increased food supply is both a prerequisite and an impetus to improved distribution. Especially in tropical Africa, where rural inhabitants comprise around 80 percent of the total population, improved income distribution cannot occur without increasing farm output. And increased farm output will not materialize unless the farmer sees the connection between increased output and an improvement in his own income.

Even the production of so-called luxury goods such as coffee and cocoa can indirectly improve the peasant's food consumption by increasing his available income. In tropical Africa this is true even for many of the poorer peasants because access to income-producing assets such as cocoa trees is not limited to large plantation owners. Indeed, the bulk of production of such products has often been derived from the small farmer in the past.[12] The failure of the tropical African farmer to take advantage of production potentials has been due more to privileges and policies generated by regimes intent upon industrializing or bolstering government controls over the economy than to unfair market operations. Political hierarchies, restrictions on credit and other income-producing assets, land use policies, state taxation, and agricultural procurement policies have been the primary generators of inequality. Insofar as they have stifled the ability of the African farmer to increase his production, or prevented him from reaping the benefits of increased production, such policies have significantly contributed to tropical Africa's food problem. Redistribution schemes merely attack one of the symptoms of the problem while neglecting to address its fundamental cause.

In addition to the theme citing maldistribution of assets and income as the source of tropical Africa's food problem is the argument that Africa is too dependent on the West due to an overemphasis on agricultural export crop production. Although the dependency argument is a popular one, dependency is not the cause of tropical Africa's food problem. In the first place, avoiding export specialization is not necessarily desirable, especially where there are direct natural advantages for a given product.[13] This is all the more true in tropical Africa, where export production rarely competes directly with domestic food-staple production. The export market usually exists alongside trade in domestic staples. Far from stifling domestic food output, the trade in exports has traditionally even served as a source of credit for the smaller market in food staples:

. . . the principal fact to be observed here is that the supply and the effective demand of many commodities may sometimes be too small to warrant the costs of marketing them unless they can be handled as sidelines. Measures that either cut private merchants off from the large-volume trades or that prevent merchants licensed to engage in a large-volume trade from dealing in other commodities inevitably inhibit the development of new markets and new production for sale.[14]

Other side effects of the export market similarly stimulate the domestic foodstuffs market. Infrastructure facilities that develop around the export market including transportation, information services, and so on serve as a basis from which internal markets can also develop. These facilities help to regularize the flow of supplies to the market, thus encouraging demand. As individuals become increasingly assured of being able to buy food in the market they, in turn, can move from reliance on subsistence production to specialization.

Fact and Fancy regarding Tropical Africa's Food Problem

A plethora of propositions has been offered to explain the unimpressive gains in tropical African farm output, pointing to everything from physical to cultural to political constraints as the culprit. The cultural theme has several components. First, the African farmer is considered irrational or improvident, lazy and resistant to change. A more sophisticated criticism of African society blames tribal isolation, fragmentation of farm plots, and the importance of the extended family. This last factor, it is claimed, is responsible for immersing the African farmer in a network of family obligations that make it impossible to accumulate a stock of goods for trade. Finally, even where market ventures do emerge, kinship ties result in a redundant number of family members being employed as middlemen, thereby increasing market costs and diverting labor from more productive activities.[15]

None of the recent efforts to analyze the operations of the African small farmer support these criticisms of African society. One African expert provides ample testimony of rational farmers responding with considerable initiative to market incentives. She notes that the irrational farmer myth is often bolstered by a failure to recognize forms of capital accumulation not generally considered in orthodox calculations of savings in developed economies.[16] Another analyst similarly contradicts the assertion that the African farmer is either irrational or lazy;

Farmers . . . are sufficiently alert to take advantage of attractive alternatives for sale when these are open to them. . . . The farmer's decision whether to sell at the time of harvest, to sell a month or more later, or to hold for consumption by his own family depends on a complex set of calculations.[17]

Nor can the claim that unduly numerous middlemen interrupt the marketing chain be substantiated. Indeed, if anything, the market chains are short. Detailed studies of trading patterns in Nigeria, Sierra Leone, and Kenya show similar results, with usually "no more than two middlemen—a distributing wholesaler and an assembler—between the grower and retailer; typically there is only one."[18] The same study indicates that there are no

institutional or trade barriers that would prevent the producer or trader from bypassing the intermediaries in the buying chain; on the other hand, they report few instances when customary intermediaries are bypassed. Even in cereal markets growers tend to sell to a retailer or wholesaler, although they are perfectly free to sell directly to consumers. Apparently the advantages that might accrue from direct sale are offset by the risk, inconvenience and the waiting that such sales involve. The major exceptions to this tendency to sell through the usual chain are found in the Sierra Leone rice trade and especially in the Kenya maize trade where there is a strong incentive to bypass the official marketing agency. There are no other indications that the number of intermediaries is excessive in terms of the length of the marketing chain.[19]

A second claim, somewhat related to the first, contends that African agricultural practices are inefficient. Preference for small-scale farming purportedly impedes specialization and a more rational allocation of scarce resources. In the long run, it is probably true that some forms of specialization will be imperative if significant gains in agricultural production are to occur. However, in the short run, with market systems relatively small, large-scale production for sale is not necessarily more efficient. Furthermore, ample evidence of efficient farming techniques among small farmers exists. One analyst contends that multicropping techniques often make good economic sense. She notes that cocoyam and plantain plantings in cocoa fields are necessary cover crops for the young seedlings, and concludes that such farming methods may be efficient where (1) labor rather than land is scarce, (2) it is cheaper to grow some foods rather than buy them considering the bulkiness (hence difficulties in transporting) of certain staple African foodstuffs such as cassava, and (3) existing land arrangements make large-scale production or mechanization inappropriate.[20] For much of the forest zones of East and West Africa the above conditions prevail, making much of the conventional wisdom concerning large-scale farming inappropriate.

A third myth regarding African agriculture maintains that inadequate market demand for foodstuffs stifles any effort toward surplus production for trade. However, ample evidence exists to suggest that small markets are not necessarily inefficient, and may indeed cheaply and regularly provide desired commodities to consumers, both justifying further demand and stimulating the increased marketing of production.[21] One study evaluates the assertion that effective demand is inadequate by investigating the magnitude of current staple food trade, the potential for expanding market volume, the degree of self-sufficiency in rural Africa, and the behavior of prices in rural markets. The study concludes that demand is generally adequate. Constraints on demand, where they exist, appear to derive not from mistrust or traditional bias, but from restraints placed on private trade and from the prevalence of low incomes.[22]

Infrastructural bottlenecks, both physical and institutional, have been popularly blamed for Africa's agricultural problems. No doubt a relative paucity of roads, storage facilities, credit arrangements, and modern technology characterizes most of tropical Africa. But the importance of these infrastructural shortcomings in the overall picture of African agricultural performance has been considerably overstated.

Countries such as Zaire, Uganda, and Ethiopia have suffered frequent interruption of more sophisticated means of transport such as roads and railways, especially as a result of wars and internal strife. Where transport problems do seriously inhibit market operations, efforts to invest in road construction might contribute to tropical Africa's agricultural development. However, road-building is generally costly and efforts to improve existing markets by removing trading and pricing restrictions are likely to be more effective. In any event, one analyst of tropical Africa's agricultural production systems notes that, when effective demand has stimulated production, private farmers themselves have actually undertaken to provide roads to facilitate bringing their goods to market. The implication is that, where improved transport facilities are desired, effective means of transporting the goods in question often spontaneously emerge.[23]

Rural market imperfections appear to be less the result of poor communications and transportation systems than of government policies systematically biased in favor of urban elites and economic centralization. Nor is there any inherent deficiency in the ability of tropical Africa's physical endowments to support increased food production competitively. Tropical Africa possesses four basic rainfall zones each facing substantially different production constraints.[24] Generally, African soils are chemically poor and are frequently subject to severe wind and water erosion problems. However, several factors weaken the contention that the general region comprising tropical Africa is unsuitable for the widespread, efficient production of staple crops. First, each area has largely been able to grow at least one staple suitable under the prevailing agricultural conditions. Sorghum, millet, maize, rice, wheat, cassava, and plantains manifest a wide array of growing requirements, which generally make at least one of them suited to any of the different growing regions of tropical Africa. Second, improved seeds, irrigation possibilities, and fertilizers all enhance the growing potential of less well endowed tropical African areas. These technical improvements are not always available to the African small farmer, nor are their costs always merited. Nonetheless, the land is susceptible to manipulation by technical improvements and the latter need not be in the form of modern, megalithic projects.

The Political Economy of African Agriculture: Planning Zeal and Urban Bias

Ultimately the poor performance of African agriculture cannot be attributed to cultural barriers, entrenched traditional farming techniques, lack of demand, physical constraints, or infrastructural defects. Instead, the key to Africa's food problem lies in the complex web of interrelationships between Africa's agricultural markets and government policies. Both markets and government policies are often overlooked—first, because they are complex, and second, because pronouncements concerning them are more politically volatile than are evaluations of technical and physical problems. But it is precisely within this more controversial arena that understanding of Africa's food problem must be sought.

THE AFRICAN FOOD MARKET

Competitive markets have been much maligned in recent years, particularly by development theorists. However, the considerable shortcoming of planning alternatives, coupled with renewed interest in market mechanisms even within planned economies, suggests that their relevance to production cannot be dismissed. For most of tropical Africa, markets, though manipulated by government policies, continue to be the primary means by which farmers are induced to increase agricultural output.[25] Rural producers and consumers keenly appreciate the advantages of marketing agricultural products over traditional subsistence production, in which little or no trading takes place. It reduces dependence on their own individual resources. It permits specialization by individual farmers so that advantage can be taken of favorable local conditions. In addition, larger markets permit increased possibilities for improving processing, storage, and transportation facilities. Finally, taking advantage of market opportunities enhances the income-earning potential of the farmer, thereby directly improving his ability to provide a nutritional diet for himself and his family.

Despair that African markets are disorganized and impossibly complex fuels arguments that potential interrelationships of supply, demand, and price are rendered inoperative. However, substantial testimony exists to the contrary.[26] Several characteristics of the tropical African market deserve particular attention since they are especially relevant to the African food problem. What is the time lapse between harvests, which might influence the need for storage? What are the costs of such storage? What are the costs of transporting goods to the market? How quickly is the farmer's decision to increase production translated into available supplies in the market, and the opposite direction, how quickly do shortages in the market influence a farmer's decision to increase planting?

Each commodity marketed in African food trade manifests markedly different qualities in relationship to the above factors. William Jones summarizes his analysis of crop characteristics by concluding that the "time lag between harvests for most of the crops is relatively short, frequently less than six months."[27] This relatively short time period between planting and harvest permits a rather rapid response to price stimulus resulting from supply shortages or increased demand. In addition, most of the major crops store well without severe loss, particularly if optimum conditions can be achieved with the aid of air circulation and fumigation. Similarly, with the exception of root crops and bananas, most African staples are easily transported at reasonable cost, which permits the movement of supplies from surplus areas to deficit areas.

Regarding the lack of uniform quality and measurement standards that impede price comparisons, one agricultural expert cautions against presuming that such factors are evidence of a chaotic, unreliable market. The creation of both types of standards often imply costs that may not be merited in a market where consumer incomes are low, and demand is distinctly limited by lack of purchasing power.[28] Efforts to impose such standards may reduce the overall effective demand for products, forcing poor families to revert to subsistence practices, or to purchase less costly and less nutritive staples than those whose quality has been controlled.

Related to the problem of price dissemination is the difficulty with which farmers in isolated areas receive price and demand signals. The farmer may be at a serious bargaining disadvantage, especially where only one buyer may venture into such isolated areas to purchase what meager marketable supplies may exist. Should he repeatedly fail to be able to capitalize on higher prices in more central markets he may abandon efforts to grow food for sale. However, the plight of the isolated farmer, while unfortunate, does not imply the overall absence of an effective market system in more central locations. Though such farmers may eventually become more closely linked to a general market system, it is in the further stimulation of already integrated market networks that the greatest immediate potential for improved production in response to increased demand lies.

One final dimension of the African agricultural market that has thus far been only alluded to is that of credit. It is often claimed that shortages of credit have inhibited the expansion of African markets, and thus of food production. Lack of credit diminishes the possibilities for economies of scale in assemblage, transport, storage, and processing to materialize. Similarly, it impedes the potential for transporting food from surplus areas to deficit areas. Perhaps most important, it denies producers the opportunity to purchase materials, including seed and fertilizer that might significantly improve their production levels.

In large part, insufficient credit is the result of government policies affecting interest rates and collateral availability. However, at least part of the problem can be attributed to prevailing economic structures. The nature of land ownership in tropical Africa presents a peculiar problem for credit provision. Elsewhere, private land ownership has often served as a basis for collateral in loan transactions. In tropical Africa, although most peasants cultivate plots without payment of rent they do not own their land and hence cannot use it as collateral for obtaining loans. Freehold ownership has begun to evolve spontaneously in parts of tropical Africa, but titled registration of ownership remains extremely limited.[29]

Land ownership may not necessarily be a prerequisite to an effective credit system if warehouse receipts, or other production assets, could be utilized as collateral. To some extent such arrangements have actually materialized in tropical Africa. But government restrictions have often impeded widespread development of such mechanisms for facilitating the provision of credit, thus profoundly illustrating the tight interconnection of economic and political institutions. Because this interrelationship is particularly highlighted concerning the question of credit availability, a more thorough discussion of the constraints and possibilities for improving credit opportunities will be set forth in conjunction with the analysis of government policy and tropical African food production.

One comment does deserve note. It has often been assumed that lack of credit reflects an overall inability or disinclination of Africans to accumulate capital. The contention reflects misinformation and a narrow understanding of the nature of capital. One intensive study of rural capitalism in Africa exposes this misconception:

> . . . the economist is so unfamiliar with the forms such saving and investment are apt to take that he does not know where to look for evidence of their existence. (I have heard UN financial experts seriously argue that the admitted unwillingness of individual Ghanaians to buy government securities is evidence of their inability to "save" in any sense of that word). . . . At present, many of the most important forms of fixed capital, such as cocoa farms, are omitted from all official national accounts, mainly because this reflects traditional accounting practice in developed countries, but also because such farms are wrongly presumed to have come into existence almost accidentally—to be acts of God, rather than manmade capital assets, the creation of which involved much effort, abstention, and planning.[30]

If one is to understand the apparent lack of credit among African small farmers one must look to other than an "uneconomic man" theory, or the poor organization of Africa's agricultural market system.

AGRICULTURAL POLICY: THE POLITICS OF ECONOMIC INTERVENTION

Agricultural markets in tropical Africa are potentially viable systems, and resource constraints do not place insuperable barriers on agricultural production. Yet production has lagged and markets have failed to expand along with increased demand for food. These failures have largely occurred because government policies have suppressed the initiative to produce and distribute agricultural products. Whether specifically designed to influence agricultural markets, or simply the result of unintended side effects from industrialization programs, the result of government policies for tropical African agriculture has been drastic. Gross inequalities in distribution of both income and assets, especially between the rural and urban sectors, have resulted largely from state taxation, credit, and wage and pricing schemes that have distorted market operations. Insofar as these policies impede farmers from realizing the full gains from their labor, incentives to engage in agricultural production for sale have diminished, thus resulting in the failure of production to keep pace with demand. Furthermore, because rural incomes are directly linked to earnings from agriculture, the ability of the rural poor to purchase nutritional foods in the market has lagged.

A zeal for planning (the "do something—don't just stand there" syndrome) has generated a proliferation of policies even where information is severely lacking. Planners seem to operate on the assumption that some figures are better than none, making it likely that government interventions into food marketing will harm rather than improve the functioning of the system. Furthermore, even the availability of information in itself cannot overcome the fact that value judgments and choices are made in all government policies. This has been no less true in tropical Africa, where development choices have generally reflected a distinct urban bias. Policy motivations have been more political than economic, reflecting the expediency of responding to urban elites who are more visible even if they are less numerous than rural farmers.

A related, though somewhat different, bias against traders and merchants has also contributed to the formulation of policies whose net effect has been to suppress the marketing of agricultural production. Whether from political self-interest or a genuine desire to further the "general welfare," overall plans, government marketing boards, and licensing and credit arrangements have gained favor over the role of the individual trader. The net result has been to impede transactions that have been undertaken and expedited by such individual traders in the past.

Agricultural policy, whether the result of expediency, overzealous planning, or misinformation, has taken a number of different forms throughout tropical Africa. It is not suggested that all tropical African nations manifest identical agricultural policies. Nor does each nation's production suffer identically. Nonetheless, a bias toward certain types of policies constitutes a major constraint to improved production throughout most of tropical Africa. A look at the more prevalent kinds of policies affecting food output will highlight specific criticisms attributed to them. The policies briefly reviewed have been divided into two categories according to their most significant point of impact on Africa's markets. First, a look at forms of government interventions affecting food supply shall be assessed, followed by a review of those schemes influencing the nature of consumer demand for agricultural products.

THE IMPACT OF AGRICULTURAL POLICIES ON SUPPLY

Since bankrupt farmers are unlikely to have either the inclinination or the ability to feed a hungry world, it is remarkable that much of tropical African food policy seems to overlook this essential point. Among those policies whose net effect is precisely a reduction in overall gains for farmers the following are particularly counterproductive: price controls, licensing restrictions, credit controls, and the establishment of monopolistic state marketing boards. Of a somewhat different nature, but engendering similar disincentives to the farmer, are policies that impose cooperatives on unwilling peasants.

Price controls of one sort or another have proliferated throughout much of tropical Africa.[31] A range of widely varing justifications for such controls have been advanced. In large part, the desire for maximum price ceilings comes from urban consumers, who represent a relatively organized and vocal pressure on politicians, particularly in comparison with the less educated, poorer, and more isolated rural populations. Increases in food prices constitute a major source of discontent for urban consumers, and the threat of political unrest should food prices rise noticeably has stirred many governments to impose price ceilings on food. Juxtaposed to ceiling prices has been the implementation of schemes in which the government serves as a buyer of last resort at a fixed minimum price. The logic of these schemes has been both to stabilize prices to the farmer and to regularize supplies. Year round fixed prices, as well as price averaging schemes have also arisen with the hope of diminishing price and supply uncertainties.

Because the forms of price control vary enormously, generalization regarding their impact on production can only be impressionistically provided. However, several general comments apply to each of the schemes. First,

Governments attempting to control prices, for whatever reason, operate with some concept, however rudimentary, of the price elasticity of supply and of the desired level of supply. We have already seen the difficulties in the way of measuring price elasticities of supply for products sold on domestic markets in Africa, since few governments have information about the areas sown to the principal crops, the pattern of yields, the level of farm consumption or of local sales in village markets.[32]

Second, regardless of the specific nature of the price controls implemented one risks subsidizing inefficient farming operations, or forcing farm activity to shift from regulated to unregulated products in search of better returns. Either way, the result may be to encourage a waste of scarce resources. Third, the costs of operating any price control scheme are considerable and are often borne by either the producer or the consumer, thus inhibiting production or effective demand (which, in turn, may suppress increases in production). Finally, inadequate information and cumbersome implementation networks contribute to the supply and demand problems generated by price controls. However, it must be underlined that, even with improved information and implementation systems, price controls tend to aggravate market disequilibria, engendering either shortages or surpluses that benefit neither producer nor consumer in the long run. To illustrate this latter point, one need only look at the effects of fixed pricing program imposed to overcome price fluctuations due to seasonal supply variations on the meat market in Kenya. The outcome has been, instead, to exacerbate that fluctuation by acting as a disincentive for graziers to continue to fatten their cattle after the availability of cheap fodder during the rainy season with serious shortages subsequently ensuing. Examples in crop production similarly affected by price controls abound, particularly in Zambia,Kenya, Tanzania, and Uganda.[33]

A second category of policies that have constrained production and marketing opportunities is licensing controls on traders. Restricting trade to licensed operators derives from the assumption that free competition among traders leads to high costs, inefficient marketing, the absence of quality standardization, and potential fraud. By reducing the number of traders through licensing systems it is presumed that quality control can be more easily maintained, that economies of scale in credit, transport, storage, and processing can be realized, and that absolute costs of production can be reduced. This latter assumption derives from the belief that "each buyer and seller in a trade obtained a certain minimum income, irrespective of the volume of trade done by each; if the same volume of trade was conducted by fewer people each obtaining the same average income, total costs would fall . . . [however] they are not likely to cut this margin by half because the volume of trade has doubled for each seller, while the force of competition has been weakened."[34]

Other problems accompany efforts to license traders. Such restrictions in competition inhibit the development of commercial operations in new areas. This has been the case in Sierra Leone, where licensing of rice traders has been instituted. Licensing serves to perpetuate inefficient operations by those fortunate enough to obtain licenses yet induced to develop few improvements in their operations for want of competition.

Finally, recalling the oft-cited credit constraints in tropical African markets, licenses have often been devised with the assumption that lower numbers of traders with bigger operations would facilitate movement of credit. However, granting of licenses has frequently been accompanied by restrictions on the kind of trade in which the licensee can engage. Export traders are prohibited from engaging in the marketing of domestic foodstuffs. Yet it has been noted that it is precisely in the linking of these two markets that much traditional private credit for the domestic market has been generated.

Shortages of credit constitute a significant impediment to both improving agricultural productivity and expanding markets. Much traditional credit was derived from informal sources such as kinsmen or personal friends, or from "gold coasting," that is, export traders financing domestic production. Government policies that discourage these operations or, in the case of gold coasting, altogether prohibit it have been devised in the hope of stamping out usury and the exploitation of small farmers by lenders. However, little evidence of such abuses exists.[35] Moreover, efforts to eliminate these informal credit mechanisms have often resulted in credit being altogether unavailable to the small farmer.[36]

A second difficulty associated with the provision of credit to the agricultural sector involves the lack of collateral for many farmers. Although most tropical African farmers cultivate their land without payment of rent they do not actually own their land. Without such ownership farmers lack collateral to offer in exchange for loans. Although registered land titles are often suggested as a possible remedy to this problem, institutionalization of land ownership may not necessarily always be justified. Such registration is both costly and politically problematic where competing tribal and individual claims exist. Nonetheless, where politically feasible, such land reforms might contribute considerably to overcoming current credit problems. It might be emphasized, however, that the significance of the ownership problem would be greatly mitigated if traditional informal credit mechanisms such as gold coasting were encouraged rather than prohibited.

The widespread appearance of marketing boards in tropical Africa represents another policy constraint adversely affecting domestic food production. Their appearance dates back at least as far as World War II, when they emerged primarily as a means of accumulating capital from previous years, with the board retaining the resultant surplus. Many of these boards continued to operate after World War II as a means of controlling trade, regulating prices, enforcing grading standards, and directing foreign operations within national borders.

The operations of marketing boards, though varying widely in purpose and effectiveness, have been highly controversial. They often effectively constitute state trading monopolies that incur added costs to the consumer, reduce returns to the producer, and stifle initiative to keep costs down and provide improved services. Edith Whetham illustrates this point by citing the case of Tanzania, where a marketing board structure was introduced in the 1960s. In comparing producer and retail prices before and after the establishment of the marketing board she notes that producer prices increased only slightly while retail prices had risen by 50 percent of the earlier level, with the marketing margin more than doubling.[37]

Marketing boards have also tended to encourage black-market sales, in addition to impeding the operation of market signals regarding supply and demand. Perhaps the most uniform comment that can be made regarding marketing boards is that, as with the price controls mentioned above, insufficient knowledge about how internal food markets operate complicates the already dubious assumptions under which such policies evolve.

Government-directed cooperatives have proliferated in the past two decades as a means of promoting agricultural production by pooling rural resources, introducing modern farming techniques, and organizing remote farmers. While the formation of private cooperatives on a voluntary basis may serve important functions in encouraging the improvement of production and marketing potential in tropical Africa (especially where indigenous mutual aid traditions already operate successfully) the performance of government-imposed cooperatives to date has been unimpressive for several reasons. First, cooperatives imposed from above, with little or no knowledge of local marketing and growing traditions, have generally failed to generate the advantages they are presumed to permit. Without the enthusiastic participation of the supposed beneficiaries of cooperative programs, they are doomed to abuse and failure. Secondly, operating costs, especially where complex organizations are imposed, have exceeded any potential gains derived from economies of scale. Furthermore, where cooperatives have been instituted while simultaneously prohibiting private operations, competition that can encourage efficiency in the cooperative's activities dissolves.[38]

The above comments would seem to suggest that if the economies of scale potentially available from the formation of cooperatives are to be realized they must be based upon efficient management and the active participation of members involved. In addition, both Rene Dumont[39] and Emmanuel Okwuosa[40] point out that the peasant must have a direct interest in the cooperative's successful operation. Schemes that foster a kind of rural civil service mentality in which the farmer is a paid employee of the state, or which follow Western-type organizational structures in which resources are pooled together and owned in common, have often reduced individual motivations to contribute. Okwuosa advocates schemes in which resources are used in common but individuality is maintained.[41] In any event, it is unlikely that any government cooperative scheme is either necessary or desirable. P. T. Bauer points out that cooperatives will survive without systematic state support if this form of organization is compatible with the particular economic conditions of a given society.[42]

GOVERNMENT POLICIES AND EFFECTIVE LEVELS OF DEMAND

The potential for increased food production represents only part of the food equation in tropical Africa. Low income levels and hence limited effective demand present another serious problem. The two facets of the food crisis are intimately interrelated, especially in tropical Africa where the lowest income levels are in rural areas. Unless rural incomes can be improved, African farmers will not be able to take advantage of opportunities to expand production. On the other hand, unless incentives to increase production exist, rural incomes will also continue to lag.

The redistributive theme outlined earlier claims that the effects of maldistribution of incomes and assets severely limits the capability of markets to eliminate the food problem. It is then argued that government policies with radically redistributive effects are essential if hunger is to be effectively eliminated. The argument fails to adequately investigate the causes and, hence, the appropriate remedies for the imbalanced distribution of incomes and assets. While castigating the effects of imperfect markets on income and asset distribution, it fails to distinguish between natural market imperfections and those which result from government interventions. Although natural, physical imperfections in tropical Africa's traditional food markets exist, by far the more drastic imperfections have resulted from government interventions into the market coupled with the institutionalization of economic privileges for the politically powerful. Thus, resolution of the rural income problem (and ultimately the tropical African food crisis) lies not in further tampering with the market economy, but rather in discarding the policies that have significantly distorted income and asset distribution. Of these policies, the following have most dramatically contributed to distorting income distribution: minimum wage policies, food price ceilings, and the imposition of obligatory hygiene and grading standards.

Minimum wage legislation has been largely responsible for fueling the pronounced rural-urban income differential apparent throughout tropical Africa. Wages in industry and government employment subjected to minimum wage requirements have risen substantially as a result of political and institutional pressures. On the other hand, rural wages continue to be determined by supply and demand conditions, rather than legislated standards, and have thus remained constant, or even declined in the last decade.[43] The result for agricultural production and consumption is twofold. First, effective demand of rural consumers has been kept low relative to urban consumers, thus impeding expansion of rural markets. Second, such policies have further promoted urban migration trends exacerbating unemployment in the cities while simultaneously reducing the number of productive farmers in rural areas. This rural-urban income differential has also stifled incentives to improve agricultural production insofar as increases in returns from selling food must rise enough to match the returns in urban employment, which have been fixed arbitrarily high.

Imposition of price ceilings on food constitutes the single most important factor inhibiting agricultural output and maintaining low farm incomes. The practice has been briefly reviewed regarding its impact on food production. The effect of such price ceilings on rural-urban income disparities should also be emphasized. R. D. and B. A. Laird have cogently summarized this impact:

Perhaps the most important force working against the increased productivity of food is urban consumer pressure demanding that governments hold down food costs. This state of affairs involves a triple irony. First, in most such nations the rural poor are a majority but they are outvoted in the making of economic policy. Second, although rural people usually are poorer than city dwellers, the policies of cheap food mean the rural poor are subsidizing the urban consumers. Third, and most crucial, even though higher food costs would very likely result from ending policies that discourage agricultural production, increased food productivity should in the long run raise the standard of living for all, e.g. the export of surplus food could help pay for imports needed by the entire population.[44]

A 1970 FAO report reiterates this contention suggesting that the reluctance to raise food prices has contributed to food shortages, which in turn have generated inevitable price rises even higher than those which might have prevailed without the original price ceilings.[45]

Obligatory hygiene and grading standards have similarly negative effects on rural purchasing power. Although their application stems from the well-intentioned desire to improve the quality of food consumed, reduce potential spread of disease from unsanitary or infected food supplies, and facilitate price comparisons, their net effect, ironically, has been to reduce effective demand. The regulation of hygiene and grading standards involves substantial costs that either reduce producers' returns or, if food prices rise accordingly, diminish the purchasing capacity for food, especially of poorer families. Furthermore, the selection of the "proper" standards raises considerable value conflicts where consumers have differing requirements. Where obligatory quality schemes have been implemented in low-income countries, ironically, "the costs of good hygiene deprive the poorer families of their necessary foods, or force them back onto less nutritious diets, [and] the community may only have exchanged one set of diseases for another, with no gain in utility."[46]

Tropical African Policies: The Best Laid Plans of Mice and Men . . .

The above assessment of policies affecting tropical African food production no doubt seems overwhelmingly critical. That criticism is largely directed at the practical problems encountered in any government attempt to engineer economic activity. Such efforts are constrained by informational shortcomings out of which two pronounced problems arise. First, one may advance policies designed to resolve problems for which the wrong cause has been identified. Second, one may not have the tools (e.g., market signals) with which to evaluate and adjust policies. Either way, economic problems are likely to be aggravated, not improved. Even where the goals of such policies are meant to enhance the welfare of the poor, it has been argued that unintended negative side effects often emerge.

Centralizing economic decisions in political bodies has several potentially negative dimensions in comparison to market systems. First, problems peculiar to specific locations or individuals are blurred as aggregate national goals attain priority. Second, centralized engineering of economic activity tends to reinforce an authoritarian perspective in which individuals become malleable components of the state.[47] Setting aside the possible philosophical objections to this contention, one must question on a practical level whether such trends are congenial to encouraging individual initiative and economic incentive. Third, focusing economic decision making under central state authority politicizes economic life, heightening rather than mitigating the possibilities for strife:

When state control over social and economic life is extensive and close the achievement and the exercise of political power become all-important. Such a situation creates widespread anxiety and concern with the processes and results of political life. . . . The stakes in the fight for political power increase and the struggle for it intensifies. Such a development enhances political tension. . . .[47]

To contend that government economic intervention exacerbates problems of political instability does not imply that market operations are immune to power struggles over who gets what. However, open markets may diffuse the process of choice over a broader range of individuals.[48] Moreover, it is unlikely that the relationships of material wealth to political power are as unidirectional as much explicitly Marxist thought contends. Many Chinese in Indonesia, and Asians in Uganda, among others, amassed considerable wealth without acquiring political clout. Perhaps political power has enabled some to appropriate wealth rather than the other way around. Though a decentralized competitive market system is by no means devoid of inequalities, it is contended that power and choice are significantly more diffused than when economic decisions are concentrated in government hands.

This argument is likely to be unconvincing to those who advocate a standardization of income distribution as the basis of social justice. But if social justice implies an improvement in living standards and individual productivity and choice, then enhancing economic incentives by permitting competitive markets to thrive is essential. Eliminating interventionist economic policies would not only encourage production but also have the effect of improving income distribution, and hence food consumption, in tropical Africa. To reiterate an earlier point, the argument for encouraging competitive markets is a radical one requiring fundamental structural changes including a reduction in the scope of government, abolition of agricultural procurement taxes, and, in some instances, land reform among other things, if tropical Africa's poor are to benefit.

What Can Be Done?

The above perspective reflects a profound skepticism of government attempts to manipulate the economy. Even modest interventionist attempts at improving information flows, mitigating transportation bottlenecks, or providing credit are not necessarily merited. Such policies imply costs whose corresponding production benefits are uncertain. For example, collection and dissemination of market information can be expensive and difficult to interpret. Furthermore, knowledge of opportunity for profit is not enough. The point is well illustrated by the tale recounted of a farmer in a remote area who heard a radio broadcast of prevailing prices in the central market. The only buyer who ventured into that isolated locale offered the farmer a price less than that quoted by the radio. When informed of this fact, the buyer responded, "Go sell it to the radio."[49]

Far more promising results are likely to be forthcoming by eliminating those policies which have intervened with adverse effects on African agricultural production and income distribution. In addition, the oft-prevailing bias against traders and merchants must be overcome. Profits have fallen into disrepute in much current economic and political theory. But profits in themselves need not be abusive. Indeed, they are at the heart of stimulating productive activity. This is no less true for the trader in foodstuffs than for the farmer.

Tropical Africa has both the physical resources and the underlying market structure from which to develop a highly productive agricultural sector. Nor need that production be confined to Western-oriented export crops alone. Emmanuel Okwuosa, among others, outlines evidence of ample scope for increasing food-staple production and trade of complementary foodstuffs among tropical African nations.[50] These potential complementarities suggest another avenue for altering existing policies. Not only must internal markets be given latitude to develop freely. Trade barriers must also be reduced to foster dynamic markets among tropical African nations, as well as internationally where possible.

The failure of tropical African governments to resolve the food problem through manipulation of agricultural markets highlights the necessity of turning to alternative approaches. The heightened attention which the food crisis received in the 1970s has given impetus to the search for such alternatives. Under the threat of continued food crises, with the political repercussions that such crises might engender, tropical African policymakers may be more susceptible to reconsidering the wisdom of interventionist policies. Insofar as resolving the food problem is a genuine goal of existing tropical African regimes, openness to a market approach is not entirely improbable. The relative success of Cameroon, which has limited its activities primarily to road-building while leaving agricultural markets to operate freely, provide concrete evidence to attract policymakers toward a similar approach. Furthermore, while the "ideal" solution may be to radically limit government economic activities and to fundamentally restructure political power configurations, incrementally removing licensing restrictions, price ceilings on agricultural products, and coercive procurement policies would considerably improve both the incentive and the capacity of farmers to increase production. In addition, even where marketing boards are retained for political reasons, permitting private traders to compete with government operations would substantially benefit both producers and consumers by making alternative sources of supply and demand available. The same might be suggested with regard to government cooperatives and government credit institutions. While it may be politically impossible to promote the abolition of such activities, it might nonetheless be possible to promote the establishment of private alternatives alongside government operations as a means of stimulating efficiency and individual incentive. In addition, rather than make grading and measurement standards obligatory, their provision could be made voluntary. Such measures would substantially expand the scope of alternatives available to tropical Africa'a farmers, creating conditions under which production can increase and rural incomes can improve. This combination's effect on production and incomes may go a long way toward resolving tropical Africa's food problems.

Unfortunately, proposing economic remedies is always easier than implementing them. Eliminating existing policies that constrain both domestic and international trade in food staples is certainly possible. However, a more politically contentious set of policies in a world apparently moving toward protectionism would be difficult to conceive. Nonetheless, to shy away from such proposals would only further divert attention from the fundamental problem by attempting to find a politically expedient temporary palliative.

Notes

1. UNFAO, *Monthly bulletin of Agricultural Economy and Statistics*, vol. 1, 1978, and vol. 2, 1979, nos. 1, 2, and 3; and UNFAO, *Production Yearbook*, vol. 28, 1974.

2. For tropical Africa's agricultural productivity potential see Emmanuel Okwuosa, *New Direction for Economic Development in Africa* (London: Africa Books, 1976), pp. 128, 154; W. D. Shrader, "Soil Resources—Characteristics, Potentials, and Limitations," in E. R. Duncan, *Dimensions of World Food Problems* (Ames, Iowa: Iowa State University Press, 1977); and S. M. Makings, *Agricultural Problems of Developing Countries in Africa* (Lusaka: Oxford University Press, 1967).

3. For a more thorough analysis of world food production and trade, see Louis Thompson, "Food-Producing Regions of the World," in Duncan, *World Food Problems*.

4. For an excellent discussion of this theme as it pertains to the world food problem, see Keith O. Campbell, *Food for the Future* (Lincoln, Neb.: University of Nebraska Press, 1979).

5. Jonathan Kapstein, "Spreading Starvation Stalks Africa," *Business Week*, 28 May 1979, p. 76.

6. Population figures are drawn from the *UN Demographic Yearbook*, 1977; *Population and Vital Statistics Report*, UN Statistical Papers, series A vol. 31, no. 2; *World Population 1977: Recent Demographic Estimates for Countries and Regions of the World*, U.S. Bureau of Census; and E. R. Duncan, *World Food Problems*.

7. William Jones, *Marketing of Staple Food Crops in Tropical Africa* (Ithaca, N.Y.: Cornell University Press, 1972) p. 34.

8. Ibid., pp. 28-30.

9. Ibid., p. 29. Edith Whetham, *Agricultural Marketing in Africa* (London: Oxford University Press, 1972) pp. 21–22, generally concurs with Jone's assessment but notes that diets of women and children may be deficient, as are urban diets, which manifest less variety than the typical rural diet even though incomes are generally higher.

10. A number of studies have investigated the relationship of employment and incomes to population growth. See, for example, G. Ohlin, *Population Control and Economic Development* (Paris, Development Centre of the Organization for Economic Co-operation and Development, 1967); S. Enhe, "The Economic Aspects of Slowing Population Growth," *Economic Journal*, vol. 76, no. 301 (March 1966); T. Paul Schultz, *Fertility Determinants: A Theory, Evidence, and an Application in Policy Evaluation* (Santa Monica, Calif.: Rand, 1974).

11. Cheryl Christiansen, "World Hunger: A Structural Approach," *International Organization, vol. 32, no. 3 (Summer 1978)*.

12. Polly Hill, *Rural Capitalism in West Africa* (Cambridge: Cambridge University Press, 1970), ch. 2.

13. Rene Dumont, *African Agricultural Development* (New York: UNFAO, 1966), p. 45.

14. Jones, *Marketing of Staple Food Crops*, pp. 251–52.

15. See, for example, Guy Hunter, *The New Societies of Tropical Africa: A Selective Study* (London: Oxford University Press, 1962); and V. R. Dorjahn, "African Traders in Sierra Leone," in *Markets in Africa*, P. J. Bohannan and George Dalton (Evanston, Ill.: Northwestern University Press, 1962). See also Laura and P. J. Bohannan, *Tiv Economy* (Evanston, Ill.: Northwestern University Press, 1968). Comments to this effect by local African government officials are also cited in Jones, *Marketing of Staple Food Crops*, p. 10.

16. Hill, *Rural Capitalism*, p. 11.

17. Jones, *Marketing of Staple Food Crops*, pp. 241–42.

18. Ibid., pp. 90, 176, 215, and 239.

19. Ibid., p. 239.

20. Hill, *Rural Capitalism*, p. 16.

21. Whetham, *Agricultural Marketing*, p. 35.

22. This theme is repeatedly reaffirmed in the studies by Jones, Hill, and Whetham cited earlier.

23. Polly Hill, *The Migrant Cocoa Farmers of Southern Ghana: A Study in Rural Capitalism* (Cambridge: Cambridge University Press, 1963).

24. M. K. Bennet, "An Agroclimatic Mapping of Africa," Food Research Institute Studies, III, 1962. Bennet describes four tropical growing zones in Africa:(1) the successive crop zone, (2) the moister single-crop summer growth zone, (3) the drier single-crop summer growth zone, and (4) the rain-deficient zone. The first zone supports yam, cassava, cocoyam, rice, and oil palm production. The second zone supports primarily maize, which is also grown in the third zone. Sorghum, millet, and wheat can grow in the rain-deficient zone.

25. William Jones, Edith Whetham, and Keith Campbell extensively expound upon this theme, but it is also implicit in Polly Hill's works. Jones, p. 2, comments that "an efficiently integrated and accurately responsive market mechanism is of critical importance for optimum allocation of resources in agriculture, for stimulating farmers to increase output, and for assurance of regular and low-cost supplies of foodstuffs and raw materials to the other sectors."

26. William Jones, Polly Hill, and Edith Whetham devote extensive research toward investigating the proposition that African domestic food markets are inefficient and disorganized. None finds evidence of this contention. Jones, p. 272, concludes his study unequivocally, contending that "our studies have clearly demonstrated that African traders, operating through

freely organized markets, have the capacity to carry out all normal marketing functions in a reasonably adequate fashion."

27. Jones, *Marketing of Staple Food Crops*, p. 84. For a more detailed analysis of commodity characteristics in tropical Africa, see Jones, pp. 74–84. See also William Jones, *Manioc in Africa* (Stanford, Calif.: Stanford University Press, 1959); B. F. Johnston, *The Staple Food Economies of Western Tropical Africa* (Stanford, Calif.: Stanford University Press, 1958); and M. P. Miracle, *Maize in Tropical Africa* (Madison, Wis.: University of Wisconsin Press, 1966).

28. Whetham, *Agricultural Marketing*, p. 35.

29. Ibid., preface and ch. 10. Whetham cites, for example, the emergence of freehold tenure among the Bugana, and also among the Tonga in Zambia, as well as among some Ghana cocoa farmers. However, land ownership systems are extremely diverse, depending upon local customs, market structures, etc. Mixtures of communal and individual land ownership seem most common under current conditions. Outright state ownership of land has been rather unpopular and "is limited by the strong sense of the rights of the occupiers holding land by a variety of customs," p. 201. See also Carl Eicher, *Research on Agricultural Development in Five English-Speaking Countries in West Africa* (New York: Agricultural Development Council, 1970); Victor Uchendu, "Some Issues in African Land Tenure," *Tropical Agriculture*, vol. 44, no. 2 1967.

30. Hill, *Rural Capitalism*, p. 11.

31. For a deatiled analysis of such price controls, see Whetham, *Agricultural Marketing*, ch. 9, and L. E. Preston, "Market Controls in Developing Countries," in *Journal of Development Studies*, vol. 4, no. 4 1968.

32. Whetham, *Agricultural Marketing*, pp. 183–84.

33. S.M. Makings, Edith Whetham, and William Jones each devote considerable attention to assessing the impact of various price controls on African agriculture. See, in particular, Whetham's assessment of maize pricing in Zambia, Kenya and Tanzania, of general controls in Uganda, chapters 7 and 9.

34. Whetham, *Agricultural Marketing*, p. 78.

35. For a detailed discussion of private lending and interest rates see Jones, *Marketing of Staple Food Crops*, pp. 137–38, 181.

36. Ibid., pp. 252–57; and Okwuosa, *New Direction*, pp. 148–50.

37. Whetham, *Agricultural Marketing*, pp. 132–33.

38. Rene Dumont, for example, notes the beneficial effects of competition of cooperative operations. See also K. O. Campbell, *Food for the Future*, pp. 105–7; and Whetham, *Agricultural Marketing*, pp. 188–92.

39. Dumont, *African Agricultural Development*, p. 58.

40. Okwuosa, *New Direction*, 147–53.

41. Ibid., p. 153.

42. P. T. Bauer, *Dissent on Development* (Cambridge: Harvard University Press, 1972), p. 211.

43. Eicher, *Research on Agricultural Development*, p. 23. For a specific case study of this problem, see Peter Kilby, "Industrial Relations and Wage Determination in Nigeria," in *Journal of Developing Areas*, vol. 1, no. 4 (July 1967).

44. Roy and Betty Laird, "Food Policies of Governments," in Duncan, *World Food Problems*, p. 244.

45. The conclusions of this work cited in Laird and Laird, "Food Policies of Governments," pp. 244–45.

46. Whetham, *Agricultural Marketing*, p. 35.

47. Bauer, *Dissent*, pp. 74, 84.

48. Ibid., pp. 86–87.

49. Jones, *Marketing of Staple Food Crops*, p. 260.

50. Okwuosa, *New Direction*, pp. 103–7. Okwuosa notes that Dahomey "is more suitable for maize production . . .; wheat can be grown in the highlands of Kenya and Ethiopia for export to the neighboring countries in lower latitudes; the rainfall regime in southern Ghana, Togo and Dahomey does not favor rain-fed rice cultivation compared with the areas in similar latitudes in Nigeria, Ivory Coast, Liberia, and Sierra Leone."

Part III
International Organizations

9

International Organizations and the Improbability of a Global Food Regime

SETH B. THOMPSON

At first glance it appears that there is a well-developed and articulated set of institutions comprising a global food regime at the international level. By one count, there are 89 intergovernmental organizations dealing with food in some fashion[1], with the FAO (Food and Agricultural Organization) serving as the centerpiece. There have been two World Food Congresses (1963 and 1970) and a World Food Conference (Rome, 1974). In any given year there are hundreds of meetings focused on the specifically international aspects of food issues involving national delegates and private citizens, many of them experts in a particular aspect of agriculture,[2] extensive consultations between governments in a variety of forums, and almost 7,000 international civil servants working in the FAO alone.[3]

But if central to the idea of a "regime," as Hopkins and Puchala define it, is the notion of governing, i.e., provisions to "control, regulate or otherwise lend order, continuity or predictability[4]" then it is not altogether clear that this buzz of activity deserves the label. Morever, it is most unlikely that a truly global regime, institutionalized at the international level, can be expected to emerge in the foreseeable future.

This chapter will begin by sketching some of the institutional arrangements that exist for consideration of food problems, particularly those clus-

I would like to acknowledge the assistance of Linda Teel in preparing this chapter and thank David Balaam, Michael Carey, Allen Gray, Suzanne Thompson, and David Williams for their comments. None of them are responsible for the remaining difficulties, and all are probably entitled to say "I told you so."

tered around the FAO, and will describe the strategies adopted by the FAO for coming to grips with food issues and problems. Since strategies are intended to produce results, we can then ask if the FAO and other organizations have made a dent in the continuing difficulties faced by global agriculture.

Put simply, the answer is "no," if the criterion is change in the incidence of malnutrition or increases in food security for the vulnerable. Even if the criterion is more modest, e.g., creation of a stable and predictable global market, the answer remains negative. It is not so much that the activity since 1945 has been altogether fruitless (among other things, the amount of hard data and analysis on a large number of specific issues in world agriculture has increased tremendously as a result of FAO studies) as it is that the activity has been irrelevant to the problems of altering the central dynamics of the political and economic systems involved with world food issues.

The causes of this outcome have not been so much a lack of effort on the part of the FAO and other organizations, nor a lack of diligence in trying to identify key factors and dynamics, although food organizations have not been exceptions to the pattern of bureaucratic shortcomings identified with contemporary international organizations.[5] The roots of impotence lie instead in (1) the marginal position of international institutions to the world food system, and (2) the existence of a series of conceptual/political problems that have blocked the emergence of a coherent strategy that can be pursued over time. Since it is improbable that either of these sets of factors will be altered very soon, it is improbable that a global food regime centered around international organizations will emerge.

Form, Function, and Growth

The purpose of this section is not to provide a comprehensive overview of organizations active in the international politics of food but to provide a suggestive description of some of the highlights. International institutions concerned with food are not latecomers to the scene. As early as 1943, a United Nations Conference on Food and Agriculture was convened in Hot Springs, Virginia, at the behest of the United States to consider the postwar organization of food supplies. The conference recommended creation of a permanent body. This was effected in October 1945, with the first session of the FAO.[6] Structurally, the organization is typical of U.N. specialized agencies with a biannual general meeting (the Conference), a smaller executive body (the Council), and a Director-General presiding over a staff (totaling over 6500 by 1976).[7] In addition to the usual relationship with ECOSOC and the General Assembly, the FAO has established a variety of working relationships with other organizations, including WHO, UNESCO, WMO, IAEA, and IMCO.

Internally, the FAO has created an array of permanent committees, commissions, panels of experts, and working groups focusing on both general and specific issues. The organization has always defined its area of competence broadly, as reflected by major standing committees on commodity problems, fisheries, forestry, agriculture, and world food security. Within this structure, some bodies have focused on particular commodities (e.g., the International Rice Commission) while others have addressed problems within a particular geographic area (e.g., the Indian Ocean Fishery Commission). The result is a near-labyrinthine set of bodies of varying size, format, and scope, often reflecting the operational complexity of a seemingly simple problem. For example, attempts to deal with plagues of desert locusts fall under the purview of the FAO Desert Locust Control Committee with its attendant Commission for Controlling the Desert Locust in the Near East, Commission to Control the Desert Locust in the Eastern Region of its Distribution, Commission to Control the Desert Locust in Northwest Africa, and a Panel of Experts on Emergency Action against the Desert Locust. (Lest a campaign against an insect seem tangential to the problems and politics of world food, it should be noted that a swarm of locusts in Ethiopia in 1958 ate 167,000 tons of cereals; enough to feed a million people for a year. Moreover, due primarily to the disruptions of the current war in Ethiopia and unusually heavy rain in the Arabian desert, the swarms have returned with a vengeance.)[8]

The FAO began to be transformed from an institution devoted almost exclusively to research and discussion into a center for field operations as early as 1950, when it was linked to the Expanded Program of Technical Assistance.[9] A major shift came with the arrival of B. R. Sen as Director-General in 1957. In 1958, Sen proposed a Freedom from Hunger Campaign under U.N. and FAO auspices to link international organizations to nongovernmental committees within member states and direct attention to nonmarket dimensions of food issues. The Campaign was launched on a five-year trial basis in 1960, expanded in 1965, and extended through 1980, despite a debate in 1963 that pitted developed nations and food exporters against less developed nations and food importers. The key issue was a fear that the program would interfere in some fashion with grain markets.[10] The thrust of the Campaign has been less to break new ground than to give focus and direction to ongoing activities and increase coordination between international, governmental, and nongovernmental groups, with an exclusive reliance on voluntary contributions.[11]

Sen also pioneered the World Food Program, a joint FAO/U.N. project. which since 1963 has used voluntary pledges of commodities, cash, and services to sponsor development programs, normally using food as partial payment for labor.[12] The WFP budget has grown from a pledged target of $84 million for 1963–65 to a target of $750 million for 1977–78.[13]

The World Food Conference of 1974[14] added two major elements to the FAO/U.N. structure: the World Food Council and the International Fund for Agricultural Development. The first is intended as a high-level policy-making body, the second as a voluntary fund (with heavy OPEC participation). Yet another body was established in 1975 when the World Bank, the FAO, and the U.N. Development Program agreed to coordinate efforts via the Consultative Group on Food Production and Investment.[15]

This activity has been reflected in the FAO regular budget, which has grown from $5 million in each of its first three years of operation to $167 million for the 1977–78 biennium;[16] in the same period membership has grown from 47 to 138.[17] In addition, FAO is responsible for the disbursement of approximately 30 percent of UNDP funds in a given year.[18]

As other chapters in this book have outlined, there have been a number of regional or subregional arrangements among governments in the area of food policy and some arrangements falling outside the scope of the FAO system, such as the International Wheat Agreements. Organizations dealing with food have been a growth industry.

"If at First You Don't Succeed . . ."

This proliferation has had a number of sources, but the primary one has been shifts in perceptions of the definition of "food problems" and corresponding shifts in international strategies for dealing with them.

In the immediate aftermath of World War II, the major problem seemed to be the restoration of prewar food trade and the provision of food to war-torn areas. At the third FAO Conference in 1947, delegates were presented with two strategic alternatives for a global food regime. Option one would have set up a World Food Board "with financial resources and wide powers to fix a general policy incumbent on the various member states" while option two looked to consultations and the preservation of national policy discretion.[19] With the understatement typical of U.N. documents, the Conference Report notes that "the latter system was preferred,"[20] which reflects not only the obvious political question of yielding sovereignty, but also the perception that the dominant problem was regularizing markets and handling surpluses in exporting countries.

The advent of a large and growing number of newly independent states into the U.N. system forced a different set of issues onto the food agenda, with increased stress on the problems of food deficit nations and the effects of global poverty. In addition, some of the most severe effects of the war had been overcome. (The 1955 FAO Conference struck a self-congratulatory tone in noting that supplies had returned to prewar levels despite a 25 percent population growth.)[21] The Freedom from Hunger Campaign was initially launched as a short-term device for focusing attention on regional disparities and distributional problems and mobilizing nongovernmental support for development efforts.

Even as the Freedom from Hunger Campaign gathered steam and began to move beyond the first step of establishing a network of national committees, it was clear that the problem of distribution remained severe. It was also clear that steps needed to be taken to increase production in less developed countries (although this was less obvious to the United States and other food exporters who feared potential market disruption). The fundamental strategy has been to supplement national development budgets by the provision of food resources to be used in lieu of cash payments to workers on infrastructure projects. The Program has been constrained by (1) reliance on voluntary pledges by governments, which has made sustained budgeting precarious, and (2) fluctuations in commodity prices. Although pledges come in the form of both cash and kind, all pledges are translated into dollars, a procedure that posed major difficulties in the period from 1971 to 1974. In the aftermath of global recession, actual contributions plummeted from $328 million in 1969 to $85 million from 1971 to 1972[22] (and 1972 dollars were worth less than 1969 dollars). Simultaneously, grain prices tripled between 1972 and 1974,[23] which meant that dwindling resources bought even less food for ultimate distribution.

The FAO has always seen itself as deeply involved in increasing the amount of information available about food supplies, trade, and consumption. In an attempt to move beyond narrow research reports and generate the kind of data that would permit global coordination and planning, the attention of the FAO staff in 1965 was focused on a long-term effort to draft an "Indicative World Plan for Agricultural Development." Almost immediately the proposal ran into difficulties, reflecting the unresolved issues that will be explored more fully in a subsequent section. Debate at the 1965 FAO Conference focused on whether the plan should concentrate on the international or national level, whether commodities or geography should be the units of analysis, and the authority of a "plan" over national decision makers.[24]

By the time the plan was in draft form in 1969, not only did the FAO have a new director-general (A. H. Boerma replaced Sen in 1968), but it was described as not so much a plan as a detailed analysis of various problems. Nonetheless, a host of objections ranging from ideological assumptions to methodological issues were raised and the document was returned to the secretariat to be essentially redone, cast now as a "Perspective Study on World Agricultural Development," a far more modest enterprise.[25] Moreover, instead of someday emerging as a final document, the study was conceived as a continuing process of analysis.

Both the Freedom from Hunger Campaign and the World Food Program presupposed an environment of production surplus in which the basic problems were fashioning distributive mechanisms without upsetting commercial markets. Generally poor harvests and significant Soviet buying altered the dimensions of the food equation in the early 1970s and created a felt need for a different strategic response from international institutions. In addition, many members were questioning the utility and relevance of the FAO itself.[26]

The FAO had already sponsored two World Food Congresses that had given some publicity to the work of the organization and the problems of food but had relatively little impact outside the FAO context. The World Food Conference of 1974 was initiated outside FAO channels, in part to attract a different set of actors (the majority of members send their Minister of Agriculture to FAO sessions; the Conference attracted Foreign Ministers). There was also an obvious intent for some serious reconstruction of the agencies involved in food issues, reflecting perceptions of both hardening bureaucratic arteries and political bias.[27]

The Conference produced both the World Food Council and the International Fund for Agricultural Development and, for better or worse, a decision to maintain much of the institutional status quo, thus reinforcing the FAO's position as the presumptive hub of institutions dealing with food at the international level.[28]

At the instance of then Director-General Boerma, the FAO Council, meeting hard on the heels of the World Food Conference, adopted an "International Undertaking on World Food Security" that projected the establishment of reserve stocks held in various nations as a hedge against the recurrence of global shortages. This "positive result of a delicately balanced compromise of many different viewpoints"[29] left implementation to individual nations (and the vagaries of their responses to internal demands and pressures) and was cast as a "pledge based on mutual trust and good faith" rather than a binding agreement or convention.[30] In addition to the predictable opposition from market-oriented forces, the easing of supply problems in the recent past has delayed implementation of the strategy by most food-exporting countries.

More recent developments in FAO approaches, especially under current Director-General Saouma, have represented not so much shifts in broad strategy as differences in emphasis, particularly a declining commitment to long-range, theoretical endeavors and increased emphasis on immediately useful technical assistance and development programming.[31]

In general, then, FAO strategy can be seen as a response to perceived needs on the part of the membership. Prior to 1950, the major problems were seen as restoring productivity and making supplies available to the areas devastated by World War II. As surpluses built up in international markets the emphasis shifted to consultations between exporting and importing nations, along with attention to the problems of less developed areas. The Freedom from Hunger Campaign and the World Food Program are both efforts to supplement commercial transactions (and their impact on food distribution) with a concessional approach and the use of food as an instrument of multilateral aid. The (apparently) short-term crises of the early 1970s saw an emphasis on food security and still more development aid, with additional structural innovations, but little change in basic approach. Given the structural and political situation discussed in the final section of this paper, this should not be surprising.

La Plus Ça Change . . .

Have the FAO and other institutional actors at the international level made a significant impact on world food problems and issues? Well,

Governments know today, as their representatives repeatedly stressed at the FAO Conference, that they can do much to raise nutritional levels even in the most disadvantaged countries. They can improve agricultural production, even in regions of adverse climate and rural overpopulation, especially if industries are developed which offer new employment opportunities. They can better the management of their forests and fisheries. They can work together in expanding and ordering their international trade. They can set about eliminating the disabilities of rural life.[32]

But that statement was drafted in 1945.

In 1974 the FAO Council reached a consensus on the current state of food and agriculture involving four points: documentation was improving on both international trade and at the national level; cereal stocks were in a dangerously depleted condition, totaling less than 50 percent of the reserves held three years earlier (which had been only marginally adequate); there was a clear overall decline in production; fertilizer shortages continued; and the members could agree to disagree about whether distributional problems could be handled by market mechanisms.[33]

In 1977, the Deputy Director-General of the FAO could report that: production was up in both more and less developed countries; supply remained heavily contingent on the weather; import deficits in the Third World were still rising as production gains were concentrated in a few countries; external assistance levels were declining and the terms of concessional grants and loans were worsening; and the number of malnourished people had probably increased. Perhaps worse, the food deficit of developing countries continues to grow, with estimates for 1985 running anywhere between 50 and 85 million tons.[34]

In general terms, it is hard to detect a significant impact for the FAO or other agencies in terms of bread-and-butter issues like production, distribution, prices, or consumption patterns. Production remains hostage to the vagaries of climate, national and regional policies on prices, and expectations about what the market will bear. Distribution remains a function of market conditions, relative wealth and poverty within a country, and national policies. Price is determined by the combination of market forces and national strategies. And consumption hinges on the foregoing factors as well as custom, culture, and tradition.

The World Food Program has had a mixed impact but has not succeeded in providing significant food security. Ideally, either the World Food Program or some other mechanism would work in countercyclical fashion, increasing aid and concessional sales when the commercial market contracts or prices escalate. Instead, the Program has worked to amplify cyclic swings, particularly in the period 1972–74 when shortages were prevalent. Not only did contributions decline, but the run up in prices meant that far fewer tons of food could actually be purchased and distributed.

But Hopkins and Puchala identify a set of new norms for the global food regime that they argue are not only highly desirable but also show some signs of emerging, including:

1. rural modernization,
2. nutrition as a human right,
3. international equity in food distribution,
4. global responsibility for development investment,
5. food aid as insurance rather than surplus disposal,
6. international responsibility for famine relief,
7. comprehensive information readily available, and
8. stable food markets.[35]

The FAO can take some credit for the emergence and acceptance of some of the values and perspectives Hopkins and Puchala describe, including greater awareness of the need for rural development and increased attention to global food problems. Perhaps the FAO's greatest immediate impact has been in the realm of increasing the flow of information about both food issues and transactions, although even here the impact is muted by the difficulty in collecting accurate production and market figures (hardly a unique problem, as the U.S. experience with Soviet grain purchases indicates[36]) as well as the insistence by some countries, most notably the U.S.S.R., that food information is a state secret.

The World Food Council and the International Fund for Agricultural Development can be seen as attempts to implement some of the emergent norms that Hopkins and Puchala posit as necessary, particularly those dealing with development and rural modernization, but they have yet to make any significant impact and the prognosis is doubtful in at least some quarters. For example, M. Ghedira, Vice President of the International Federation of Agricultural Producers, speaking from the perspective of the "family farm," told the 1977 FAO Conference that "farm organizations have had to realize that the 1974 resolutions [of the World Food Conference] were nothing more than paper resolutions; as if to prove this, the same governments adopted exactly the same resolutions in Manila last June [at the second session of the World Food Council] since over the past three years they have still not implemented them."[37] In particular, the expansion of production in 1974–76 had resulted in a price-depressing surplus rather than the creation of national grain reserves and merely reinforced producers' suspicions that "market stability" was a synonym for low prices.[38]

The Improbability of a Global Regime

How can we account for the large distance between promise and performance? Are the answers to be sought in further institutional tinkering or are there some more fundamental obstacles?[39]

Hopkins and Puchala cite both governance and norms as central to the existence of a "regime."[40] The current set of international institutions, or any plausible reconstructions, are an improbable source of governance because they are marginal to the key actors, both in terms of potential power of influence and in terms of ability to participate in concrete transactions. International institutions cannot be looked to as a source of operational norms (as opposed to general statements of preference) because there are unresolved issues embedded in the discussion of useful approaches to food issues that lead to potentially incompatible prescriptions for allocation of resources, preferences in strategy, and even images of the ideal global regime. Specifically, there is no international consensus on (1) the priority of long- versus short-term goals and (2) global versus subglobal approaches.

Global institutions are marginal in their ability to directly influence national decisions on food issues. The seeds of a globally organized food regime have been cast on the stony soil of sovereignty.

. . . the World Food Conference set about proliferating new food institutions, as constitutionally impotent as ever, but with the explanation that the older ones don't work. Needless to say, if some institutions could work, propensities to pathological behavior in the global food system might be controlled. But, then, the pathologies result, after all, precisely from the fact that the current regime is distinctly inhospitable to the notion of "working international organizations."[41]

One source of inhospitality is that at least some of the things that appear necessary from the international level intrude directly on domestic economic and development policies, for both the more developed and the less developed nations. The United States has not been alone in viewing the basic "food problem" as the management of surpluses to procure a stable and high rate of return on farm investments. Schemes for food security thrust additional values and variables into national political equations.

For less developed nations, discussions of food issues tend often to focus attention on national plans for development and the allocation of resources between competing sectors. In an address to the Economic and Social Council, then Director-General of the FAO Boerma pointed out that constraints on agriculture are constraints on development and went on to identify the most important constraints. Weather topped the list, followed closely by the failure of governments to "accord sufficient priority to agriculture and rural areas in general. More than any other sector, agriculture has suffered from a lack of sustained political will and commitment on the part of governments."[42] Land tenure systems and poverty were also high on Boerma's list; both are factors intimately related to the stability, if not the survival, of any Third World regimes.

Writing from a promarket perspective, Abdullah Saleh lists nine major disincentives to agricultural productivity prevalent in less developed countries, ranging from controls on prices paid to producers and by consumers, to subsidized imports, to restrictions on the internal movement of commodities.[43] For the most part, these are not accidents but the consequences of national policies adopted in pursuit of particular internal interests.[44] In late 1976 the Egyptian government announced a sharp reduction in the subsidy levels for basic foodstuffs in urban areas; the resulting riots directly threatened the Sadat regime.

Global dimensions of food issues intersect and interact with domestic dimensions; it strains credulity to think that governments are on the verge of ceding a significant measure of control over their economic life to any other institutions. Moreover, the current performance of international institutions probably has political utility for states, as a mode of symbolic reassurance.[45] Governments can direct attention to the inadequacies of global policymaking institutions to divert attention from their own policies and their consequences. A World Food Conference is a necessarily rare event, cast as a major confrontation between diverse views and competing priorities, an exercise in "high politics." But unless it results in concrete action, it looks more like a typical exercise in conceding and compromising on symbolic positions than a harbinger of changes in resource allocation or patterns of behavior. If at least some less developed nations can escalate food issues to the international level to deflect internal pressures, at least some more developed nations can offer symbolic concessions in lieu of action to their Third World critics.

Even if nations had the "political will" (however one might operationalize that term) to establish governing organs for a global food system, international institutions would remain marginal to the overwhelming bulk of the human and social transactions that can be said to make up a world food system. In the recent past, no more than 10 percent to 12 percent of world grain production has moved internationally, as compared with 50 percent of oil production.[46] In food aid and concessional sales, bilateral rather than multilateral arrangements dominate. In 1974 U.S. concessional sales under PL 480 dropped below 41 billion for the first time since the initiation of the program;[47] the World Food Program had a target of $440 million in pledges for 1974–75.[48] State behavior clearly reflects a pattern of behavior "decentralized into national sub-systems where most production, consumption and exchange take place."[49]

The experience of various governments strongly suggests that even national decision makers are often marginal to basic food decisions. During the 1950s and 1960s the United States government attempted to coordinate and control farm markets within the country by imposing acreage restrictions to curb production. The application of increased amounts of technology per acre effectively subverted that policy. The Soviet Union has the longest experience of attempting to control decisions and behavior in agricultural production; Soviet efforts have produced, at best, indifferent results. The difficulty of governmental direction of production pales in comparison to directing consumption. Short of full-scale rationing, it is difficult to see governments playing any effective role beyond exhortation.

Food can be seen as different in kind from most "problems" that fill the agenda for international cooperation precisely because there are few key decisions or decision makers. In effect, the series of outcomes subsumed under a listing of world food problems are the cumulative result of decisions about production on the parts of hundreds of millions of farmers and decisions about consumption made daily by every person on the face of the globe. It is unlikely that any coherent regime can be imagined, let alone constructed, at the international level to provide direct impact on those decisions. The likelihood of "direction" or "control" is even lower.

Human behavior in relation to food, from the everyday rituals of preparation and consumption to the complex market operations of importers and exporters, is remarkably complex and variegated. Much of the political discussion at the international level has involved mutually contradictory positions and proposals and divergent analyses of what the "real" problems are. While states and analysts find it relatively easy to agree on at least some of the general goals to be pursued in relation to food, assigning priorities and specifying strategies involves unresolved conceptual and political issues. The issues are conceptual inasmuch as they revolve around alternative understandings of the basic dynamics of food issues and political because they have direct implications for resource allocations. While there are a large number of such issues, they can be subsumed under two general categories: the relative priority of long- versus short-term goals; and the relative utility of a global versus subglobal focus.

When Director-General Saouma took office in 1976, he announced a sweeping reorientation of the FAO and its programs. He called for a shift from long-term "theoretical studies" to "more practical programs" and marked cuts in FAO-sponsored meetings, publications, and headquarters personnel.[50] In many respects this echoed and ratified the complaints of members at the 1975 FAO Council meeting when they noted, among other things, that the staff had scheduled over 1,000 meetings of various sorts for 1976–77, an increase of 40 percent over the preceding two years. Not only was the number of meetings questioned, but the basic need for sessions of experts was challenged and a clear preference was expressed for regional meetings using national resources.[51]

Yet many of these states had been among those most critical of the 1969 version of the Indicative Plan because it paid too little attention to long-range, structural factors such as reforms in the international trading system and a coherent analysis of commercial and concessional policies.[52]

The issue would be difficult to resolve if it only involved questions of marginal returns from alternatives, i.e., does an additional dollar invested in long-term research yield a greater or smaller payoff than an additional dollar investment in technical assistance? But there is a real possibility that stressing the value of either long- or short-term projects creates unanticipated negative consequences.

Efforts to increase production in the near term, for example, particularly in the more developed nations, in the absence of mechanisms for intervening in markets (e.g., some form of coordinated stockpiling) are likely to re-create the boom-bust cycle of the 1950s and 1960s and, in fact, provide net disincentives to producers.[53] The "Green Revolution" bought a short-term increase in production in some less developed countries, with the consequence of increased dependence on imported oil and fertilizers. The long-term effects became excruciatingly clear with the post-1973 increase in oil and fertilizer prices. Moreover, at least for the moment, the most effective strategies for increasing production feature the application of capital-intensive farming methods; short-term production gains are traded off for long-term problems flowing from the migration of "surplus" rural labor into urban areas. Not the least of these problems is that swelling urban populations increase demands for low food prices, ironically decreasing farmers' incentives to take the risks necessary to expand production.

Whatever the shortcomings of a temporally narrow focus, only those who are already well fed can afford the seeming luxury of maintaining a clear perspective on the long run (when, as Keynes pointed out, we shall all be dead). A long-term strategy offers the promise of substantial future benefits at the cost of considerable suffering now.

Of course this is not an either/or question in principle. International institutions can do a little of both, as can national governments. But so long as there is not even a tentative consensus on priorities, there is little chance that international institutions can make much of a contribution to regularizing, let alone "guiding" a global system. While the World Food Council was promoted as a device for providing precisely that kind of overall guidance, the record so far suggests that delegates to that body have been no more successful than anyone else in coming to grips with the issues.[54] The prospect, then, if for continued backing and filling, juggling and altering operational priorities, as vision flickers between today and the day after tomorrow.

A second fundamental question revolves around the issue of the level of analysis and action. Is it more fruitful to conceptualize and attack food issues at the global level, or do relevant answers emerge only at the regional or national levels?

Most of the efforts of the FAO and other U.N. agencies have taken a global view of food issues. The Freedom from Hunger Campaign, the World Food Program, the Indicative Plan, and the World Food Council were all efforts at a relatively comprehensive global approach. Along the same lines, those who concentrate on international food markets and/or commodities agreements can lay claim to a global perspective. Since production of surplus food is concentrated in a few states and separated from the hungry by geographic and political barriers, there are significant advantages to looking from the top down, as it were.

Even if states were committed to eradicating hunger as a first priority, the commitment would be irrelevant in the absence of a comprehensive analysis and plan of action that treated food behavior as if it were a global system. The individual need for food is a global human constant, even if the dynamics of production and distribution are decentralized and heavily influenced by local or regional variation. In many ways the current difficulties are not the result of malevolence or greed on the part of some, but the consequences of a lack of coordination and the paucity of information on the global impacts of regional, national, and individual behavior. Policies and practices that are rational at one level aggregate into collective irrationality at another. The desirability of integrating food and commodity concerns into a comprehensive global framework is clearly recognized in those elements of the call for a New International Economic Order that deal with agriculture.[55]

And yet, we have a book promoting a regional approach to food politics. At the international level it is hard to argue for a unified system when commercial food transactions tend to be concentrated between the more developed nations of North America, Western Europe, and Japan, while concessional transactions flow north-south. Identifying world food markets and attempting to coordinate and rationalize them means defining a particular game with a given set of players (primarily major importing and exporting states), along the lines of the International Wheat Agreements. Defining the game as enhancing production and distribution within a particular geographic or cultural area identifies a different set of players and rules. "Neither approach is integrated in a coherent conceptual framework."[56] The debate at the FAO Council in 1975 over proposed guidelines for international agricultural adjustment played heavily on the loss of clarity flowing from the attempt at universality, as well as the relevance of global planning to complex regional or national situations.[57]

If the earlier comments on the structural marginality of international organizations to the basic dynamics of decision making on food issues are taken seriously, there is an air of futility to any approach cast at the rarified level of a presumptive global system.

Again, it is not an either/or situation, although it would be neater if it were. A global perspective highlights one set of problems and prospects; subglobal analysis highlight other sets. The sets are real; they have implications for action, and they call into question many current policies and practices. Global approaches often have the virtue of breadth and coherence, at the expense of oversimplification and detachment from behavioural dynamics; lower-level approaches buy clear views of the trees at the risk of losing sight of the forest.

Ultimately, the conceptual issues being sketched here represent alternative sets of values. Conceptual analysis can contribute to the clarification of the values at stake and the implications of alternatives, but choosing between value-laden alternatives is a uniquely political task. Much of the disagreement and confusion flows directly from the absence of a consensus among states not only about the shape and structure of food issues but, more fundamentally, about the shape and structure of the preferred future for the international political system. Hopkins and Puchala argue that what they identify as the old consensus on food issues disappeared in the early 1970s;[58] it is hard to see it being replaced in the near future.

Conclusions

To the extent that a major property of a presumptive "regime" is the provision of direction for behavior, it is hard to see that the phrase "global food regime" has any empirical referent. International institutions are marginal to food politics in part because nations remain unwilling to cede authority to them and in part because of the very structure and dynamics of food production, distribution, and consumption. Even if international institutions were in a position to exert significant leverage on food-related behavior, there are serious conceptual and political problems involved with choosing the directions in which leverage should be used.

Does this mean that the FAO and other components of the U.N. system are irrelevant and should simple be ignored? If they ceased to exist, would anybody (other than a few thousand unemployed international civil servants) notice their passing? Or are there some concrete steps that could and should be taken to "strengthen" them in some fashion? I suspect that international food institutions are neither irrelevant nor susceptible to reforms that will enable them to play a leading role in actually altering the politics of food in the next few years: perhaps not even before the end of the century.

Clearly the international system is in a period of profound transformation. One of the symptoms of that transformation is the perception that institutions and arrangements rooted in the world of the 1950s and 1960s no longer work. It is unlikely that we will see a set of effective institutions for food, or any other issues, until the period of transition nears an end. But, in the interim, global institutions like the World Food Council and the FAO serve two vital functions.

First of all, they provide an intelligence capability for decision makers. Gathering and analyzing statistics, developing some expertise in advising policymakers on successful strategies for dealing with local problems or wooing potential donors (a prime purpose of both the FAO staff and the Consultative Group on Food Production and Investment of the World Bank, FAO, and U.N. Development Program)[59] is not a dramatic enterprise. But if policies are to be implemented at any level that in fact increase the amount of food on peoples' tables, it is a necessary one.

Secondly, if a new global consensus does arise, it will do so only through a quasi-dialectical process of confrontation, bargaining, and compromise, replete with symbolic appeals, posturing, and rhetoric. International institutions provide an established, focused, and continuing arena within which the politics of food can be played out. Not surprisingly, the arenas themselves become issues, as the pattern of events surrounding the 1974 Rome Conference illustrate. Institutional reform and restructuring makes sense when changes can be measured against a criterion of efficiency and the issues are competing means to a given end. Disagreement about basic goals cannot be dealt with by debating alternative programs of work or reorganizing the management of agencies.

Paradoxically, states have more political need of international institutions and forums when there is no consensus than they do when there is widespread agreement. Given a solid mutual understanding of ends, the problems become managerial and expertise-oriented and the solutions can often be self-executing. If nothing else, the existence of the FAO and other bodies serves as a constant reminder to states of unfinished political business and guarantees that the issues will remain on the agenda.

The myriad of formal organizations, consultative bodies, discussion groups, and panels of experts clearly suffer from a lack of coherence, order, and efficiency. In that respect they are symptomatic of the present parlous state of international politics. Nonetheless, if they did not exist states would find it necessary to invent them, if for no other reasons than to satisfy the need for information, provide an arena of confrontation, and offer some hope of moving beyond symbolic action.

Notes

1. United States Congress, Senate, Select Committee on Nutrition and Human Needs, *The United States, FAO and World Food Politics: United States Relations with an International Food Organization*, Staff Report, 94th Congress, 2d session (Washington: GPO, June 1976), pp. 11–13.

2. Ibid.

3. Moshe Sachs, ed., *The United Nations* (New York: John Wiley & Sons, 1977), p. 139.

4. Raymond Hopkins and Donald Puchala, "Perspectives in the International Relations of Food," *International Organization* 32, no. 3 (Summer 1978): 598.

5. Thomas Weiss and Robert Jordan, "Bureaucratic Politics and the World Food Conference," *World Politics*, 28, no. 3 (April 1976): p. 28, passim.

6. Sachs, *The United Nations*, p. 138.

7. Ibid.

8. Unsigned article, *Middle East*, no. 51 (January 1979), pp. 92–93.

9. *Report of the Conference of FAO*, 6th Session, 1951 (Rome: FAO), pp. 17–18.

10. *Report of the Conference of FAO*, 12th Session, 1963 (Rome: FAO), pp. 24–25.

11. Sachs, *The United Nations*, p. 141.

12. Ibid., pp. 140–41.

13. *Report of the Conference of FAO*, 18th Session, 1975 (Rome: FAO), p. 20.

14. For a succinct discussion of the results of the Conference, see Weiss and Jordan, "Bureaucratic Politics."

15. For details, see James Austin, "Institutional Dimensions of the Malnutrition Problem", *International Organization* 32, no. 3, (Summer 1978): 819–20.

16. See the appropriate FAO Conference Reports for budgetary details.

17. However, the Soviet Union has not joined FAO, which poses significant difficulties, not least of which is inaccessibility of Soviet data.

18. Sachs, *The United Nations*, p. 140.

19. *Report of the Conference of FAO*, 3rd Session, 1947 (Rome: FAO), P. 24.

20 Ibid., p. 48.

21. *Report of the Conference of FAO*, 10th Session, 1955 (Rome: FAO), p. 10.

22. *Report of the Conference of FAO*, 17th Session, 1973 (Rome: FAO), p. 21.

23. Hopkins and Puchala, "Perspectives," p. 591.

24. *Report of the Conference of FAO*, 13th Session 1965 (Rome: FAO), pp. 44–45.

25. *Report of the Conference of FAO*, 15th Session, 1969 (Rome: FAO), pp. 22–25.

26. Weiss and Jordan, "Bureaucratic Politics," p. 425.

27. Ibid., pp. 427–28.

28. R. Savary, "Editorial," *World Agriculture* 23, nos. 2/3 (1974): 3–4.

29. *Report of the Council of FAO*, 64th Session, November 1974 (Rome: FAO), par. 46.

30. Ibid., par. 48.

31. Roy Jackson (Deputy Director-General of FAO), "The World Food Situation and the Reorientation of FAO," *World Agriculture* 24, no. 4, (1977); 4–6.

32. Lester B. Pearson, letter dated January 15, 1945 transmitting the report of the first session of the Conference of the FAO, in *Report of the Conference of FAO*, 2nd Session, 1946 (Rome: FAO), pp. 10–11.

33. See note 29.

34. Roy Jackson, "The World Food Situation," p. 5.

35. Donald Puchala and Raymond Hopkins, "Toward Innovation in the Global Food Regime," *International Organization* 32, no. 3 (Summer 1978): 862–65.

36. I. M. Destler, "United States Food Policy 1972–1976: Reconciling Domestic and International Objectives," *International Organization* 32, no. 3 (Summer 1978): 624.

37. Statement by M. Ghedira to FAO Conference, 19th Session, reprinted in *World Agriculture* 26, no. 4 (1977): 19.

38. Ibid., p. 20

39. James Austin argues that institutional reform is both feasible and likely to be effective. See note 15.

40. Hopkins and Puchala, "Perspectives," p. 598.

41. Puchala and Hopkins, "Toward Innovation," pp. 860–61.

42. Address by FAO Director-General Boerma to ECOSOC, 3 July 1975, reprinted in *World Agriculture* 24, no. 12 (1975): 5.

43. Abdullah Saleh, "Disincentives to Agriculture: A Repertory," *World Agriculture* 24, no. 12 (1975): 23–30.

44. For further discussion, see Cheryl Christensen, "World Hunger: A Structural Approach," *International Organization* 32, no. 3 (Summer 1978): 745–74.

45. Murray Edeleman, *The Symbolic Uses of Politics*, (Urbana, Ill.: University of Illinois Press, 1946), especially chapter 9.

46. Henry Nau, "The Diplomacy of World Food: Goals, Capabilities, Issues and Arenas," *International Organization* 32, no. 3 (Summer 1978): 780.

47. I. M. Destler, "United States Food Policy," p. 633.

48. *Report of the Conference of FAO*, 17th Session, 1973 (Rome: FAO), p. 41.

49. Hopkins and Puchala, "Perspectives," p. 599.

50. Roy Jackson, "The World Food Situation," p. 5.

51. *Report of the Council of FAO*, 67th Session, November 1975 (Rome: FAO), pp. 57–58.

52. See note 25.

53. M. Ghedira, Statement to FAO Conference, p. 20.

54. Unsigned editorial, "Politics before Food," *World Agriculture* 24, no. 12 (1975): 3.

55. Sartaj Aziz, "The Search for Common Ground," in Antony Dolman and Jan van Ettinger, eds., *Partners in Tomorrow* (New York: E. P. Dutton, 1978), pp. 11–19.

56. Alexander Sarris and Lance Taylor, "Cereal Stocks, Food Aid and Food Security for the Poor," *World Dvelopment* 4, no. 12 (1976): 970.

57. *Report of the Council of FAO*, 67th Session, November 1976 (Rome: FAO), pp. 27–29.

58. Hopkins and Puchala, "Perspectives," p. 604.

59. James Austin, "Institutional Dimensions," p. 819.

10

Conclusion: The Regional Approach Reconciling Food Policies and Policy Recommendations

DAVID N. BALAAM

Many factors contribute to hunger problems in all regions of the world. But national and regional food strategies that deal with a number of them may be irreconcilable with wider concerns of national wealth and power distribution. This has serious implications for solving food problems and may deter efforts to create a "new world order" in which food production and distribution problems would be more consciously addressed.

This conclusion focuses on (1) the utility of the regional approach, (2) some of the more important political-economic issues surrounding food problems in different regions of the world discussed by authors in this volume, and (3) a number of policy recommendations for decision makers at different government levels who are genuinely interested in making progress toward hunger relief.

The Utility of the Regional Approach

The regional approach has three useful components. The first is a middle-range perspective that allows the analyst to move from more detailed assessments of national structural conditions and policies—as nations constitute the most predominant unit of analysis of political behavior—to assessments of larger political-economic units like the region, a definite area that encompasses a group of nation-states. Countries within regions tend to exhibit similar food problems related to the character of their natural endowments, agriculture and food policies, and relations with one another. Similar conditions are discernable within Western Europe, Latin America, the Soviet Union and Eastern Europe, the Middle East, and Africa, while regional conditions are more diverse in North America and East and Southeast Asia. Dickens and Moore make a good case for viewing North America as a region, based on geographic conditions, economic interdependencies among its three major actors, and some semblance of cultural affiliations. David Balaam makes a similar case for East and Southeast Asia based on some of the same conditions, including cultural ties and, of course, geographic proximity. But agricultural and food policies in this area are the most diverse of any region considered.

The second useful component of the regional approach is the image of reality it portrays. It is evident from the analyses in this book that there are many different food problems. And except in the case of weather, all of them are influenced by national and regional government bodies. In one sense, the EC, COMECON, and ASEAN are halfway houses between national units that cooperate with one another to achieve some end and an institutionalized global food regime. International organizations also play a role in influencing food production and distribution. But as Seth Thompson points out, even these organizations cannot break down the predominant features of nation-states. Given nation-state predominance today, even in the face of integrative pressure, interdependencies, and a number of forces that would limit nation-state influence, it seems unlikely that political arenas in the next twenty years will be organized around anything more integrated than regional organizations that will not replace the nation-states' sovereign authority in international politics.

However, it is clear that the whole world does exhibit a food problem. When enough food is produced in the world to feed everyone on the earth between 2,500 and 3,000 calories a day, no one need go hungry, or for that matter, be malnourished. The nature of the world's one common food problem is one of distribution, not necessarily of production per se. This is certainly no great revelation.[1] But it serves to direct attention to the severity of hunger, which varies a great deal according to national and regional levels of development, natural conditions, and government action or inaction.

International political-economic conditions and activities also influence food problems. In the last decade, aid programs, concessional food sales, and outright food giveaways have been cut in proportion to the seriousness of production shortfalls or economic maladies of advanced nations and to the extent that grain exporters have used their comparative advantages for other economic or more political purposes. Most agricultural trade takes place among advanced industrialized nations—between Northern country food exporters and food importers and not between Northern food exporters and the Southern developing nations who are most in need of food supplies. Even more significant, though, domestic and foreign agricultural and food policies have been greatly politicized due to the increasing linkages between domestic and foreign policy arenas, systemic interdependence, and problems related to finite resources of the earth. Food and agriculture have taken on additional roles that are more political—for example, self-sufficiency, national independence, security, and welfare concerns.

Finally, through the regional approach it becomes evident that food problems may be difficult, if not impossible, to solve due to conflicting national and regional political-economic interests that would have to be reconciled in order to implement workable solutions to a variety of food problems. Grandiose schemes that rely on single solutions cannot permeate every level of government economic and social activity that is concerned with hunger.[2] Because of the complexity of food problems, policy recommendations must be integrated to account for the many levels of human and government behavior.

What follows is a discussion of many natural and political-economic problems discussed by authors in this volume. These issues are organized into the interrelated categories of (1) natural limitations, (2) government intervention in the agricultural sector—the domestic public-private theme, (3) foreign policies, particularly wealth and power issues related to trade and security, and finally, (4) regional institutional and organizational developments. These summaries serve as a basis for highlighting the trade-offs policymakers should consider when formulating agricultural, food, or broader economic policies, and as a basis for the policy recommendations we make.

Chapter Summaries

Table 10.1 summarizes in a few key descriptive words the natural and domestic political-economic characteristics of the regions discussed in this book.

NATURAL LIMITATIONS

A factor that contributes to hunger in every region is natural limitations. Food exporters in developed regions do not face the grave natural limitations faced by the developing regions. Usually the poorest nations in developing regions are the most susceptible to the drastic changes in weather patterns that determine production outcomes. Food shortages in East and Southeast Asia are related not only to irregular weather patterns but also to the rugged terrain that hinders transportation systems. Many of the Middle East's and the Soviet Union's agricultural problems are due to a lack of adequate moisture.

These and other natural constraints set the tone for the adoption of particular trade or domestic food strategies in most countries. For example, Laos, the PRC, the Soviet Union, and a number of Middle Eastern countries import food to cities that are isolated from internal food supplies. Weather is also one reason why the Soviets abandoned self-sufficiency policies. The PRC still prefers self-sufficiency but is dependent on good weather to ensure

Table 10.1   Natural Limitations and Government Intervention Policies

	Natural limitations	Policy character	Policy goals	Policy means
North America United States	few; recent irregular weather patterns	prefer private	support farm incomes price stability	deficiency payments, targeted price supports, reserve accumulation, land diversion
Canada	" "	public accepted	" "	deficiency payment targeted to costs of production, reserve accumulation
Mexico	arid land	mixed by region, public-private	land control, self-sufficiency	National Supply Company, agri-inputs
Western Europe	few; recent irregular weather patterns	acceptance of large public role in the market, and acceptance of market influence	balance private-public welfare, price stability	structural support programs, targeted interventions, price supports, reserve accumulation
USSR & Eastern Europe	weather vagaries, and in E. Europe, excess moisture/poor soil	heavy public emphasis, acceptance of a limited role for private, E. Europe more mixed than USSR	meet consumers demand, urban biased for regime support, more equitable distribution, E. Europe also seeks self-sufficiency in some cases, and reform Soviet model	procurement prices, official support of subsidiary enterprise, and, in E. Europe, some structural programs

Region		acceptance of state's role, prefer private	comparative advantage (new goal)	mix of taxation, licensing controls
Latin America	varied			
East & Southeast Asia				
Japan	few	mix, institutionalized private influence on public	farm income support, food security	target prices, deficiency payments
High-middle income markets	drastic limitations due to weather and topography	public predominant	rationalize agriculture, realize comparative advantage, support industry & commerce	structural support programs
Low-income market	" "	mix	regime support, comparative advantage, supplement development strategy, preserve private	procurement prices
Centrally planned economies	" "	socialist, communal	equitable distribution, self-sufficiency, modernization of agriculture	mix of procurement prices, (PRC) private plots, peasant fairs
Middle East	arid regions	mix, preserve private, public role increasing	regime support, preserve private sector, support industry	mix of incoherent, ill-planned measures
Tropical Africa	weather, desertification, tropical growing conditions	public predominant	support development strategy, accumulate surpluses, urban biased	state taxation, credit, wage and price schemes, licensing restrictions, state cooperatives

the success of its new modernization program. The Soviet Union, Eastern Europe, and the PRC are all examples of nations where unpredictable growing conditions also limit the extent to which socialist goals can be pursued and where policymakers must import food and are more dependent on international markets than they prefer to be.

The area in which natural limitations have had the greatest impact on agriculture is Africa. Although Lynn Scarlett views natural constraints as a minor part of the problem, they seem to be a powerful and limiting constraint on policymakers and farmers. No type of political behavior is likely to overcome the hostile growing conditions that pervade a good portion of the African continent. Since 1968, weather patterns have continued to change in the Sahel region and have contributed to expanded desertification, droughts, and the death of some 5 million people between 1971 and 1973. The problem recurred in 1978 and 1980. The FAO has just called for one million tons of cereal to be shipped to Africa to prevent the widespread famine that, they estimate, will touch one-fifth of the continent's population. Africa continues to be a primary target for the hunger relief of many nongovernmental groups and international organizations. Of course, Scarlett's point that government behavior is responsible for some of the recent shortage in Africa is well documented.[3] Nevertheless, African governments face overwhelming natural limitations that frustrate their attempts to produce food surpluses.

Weather patterns and natural limitations are a problem not only for the developing regions but are also a major concern to developed regions, especially to food exporters whose comparative advantage as surplus producers helps balance their international trade. Since World War II both Canada and the United States have realized tremendous political and economic benefit from the almost-perfect growing conditions that make them the world's breadbasket. But beginning in 1972, production fell off dramatically due to poor growing conditions. American grain reserves dropped to an all-time record low of a twenty-eight-day supply. Trade sales and food aid were significantly cut back in the following years for fear of another shortfall. Western Europe has also been hit with bad weather at different times, which hurt production in Northern Europe.

Certainly, a host of other factors contribute to agricultural success. But natural conditions and the limits they impose on production, especially in regions that have large populations and population growth rates, cannot be underestimated in terms of their impact on agriculture and food policies. Natural barriers make planning and development strategy formulation, implementation, and outcome predictions difficult, at the very least. Agricultural investments, technological inputs, and pricing structures, among other political economic factors, influence production but, as is the case in the Soviet Union and Eastern Europe, the Middle East, and Africa, not as acutely as natural conditions. No amount of incentives or capital investments can make agricultural products grow without complementary natural conditions.

GOVERNMENT INTERVENTION

Although the character of government intervention varies from region to region, as well as from nation to nation within regions, there are a number of similarities between purposes and means of intervention in the developed regions and in the developing regions. In the more developed nations, the thrust of food politics is entangled in government efforts to balance the demands for higher food prices by small numbers of producers against a variety of demands made by nonproducer and consumer groups that have grown in numbers and political influence as the policymaking process becomes more pluralistic.

On the other hand, food policies in developing regions are more likely to resemble efforts to balance the public with the private in accordance with the goals and values of political leaders or more centralized government-bureaucratic units, which attempt to maintain control over or guide not necessarily agriculture and food development, but the direction of national development. Due to the growth of urban populations, the influence of interest groups has increased in many LDCs. Still, policymakers are more likely to prefer to "go it alone" and not to broaden access to the policymaking process.

The developed nations, especially the surplus producers, have been able to keep food production ahead of population growth and to meet the rapid increase in demand (created by large increases in per capita incomes) via a number of producer support measures, investment in agricultural research, and modernization of agricultural production, processing, and marketing. Government programs to distribute food are seldom set up, because reliance on the market is preferred over direct distribution programs. Direct distribution programs, when utilized, are used to rid the market of surpluses or to help farmers who face depressed market prices. Overproduction, which until recently was usually labeled a "farm problem," has become a problem of another kind—namely, balancing the interplay of public and private so as to derive the most benefit for all parties.

Japan's mixture of public and private interests in the agricultural sector is a model for market economies that seek to put a floor under the vagaries of the market in order to limit the damage to which producers are susceptible in such systems. While clearly capitalistic or market-dominated, the Japanese agricultural sector reflects the dominant cultural condition of a "social agreement" to guarantee influence and participation by producer and other interest groups, in determining, with their government, their fate in the marketplace. This relationship, often inaccurately stereotyped as "Japan Inc.," is actually a form of democratic-capitalistic paternalism. Japan has gone further than the other market economies to create a public-private mixture or "bond" that results in the overproduction of a number of commodities and perpetuates a great deal of economic inefficiency in food production.

In Western Europe, public intervention in agriculture in the interest of both private and public welfare has existed since the beginning of the post–World War II revival. In the Common Market, a logic of protection of private welfare and public food security prevails. But the mix of public and private wealth and power described by Carey may limit the planned expansion of more food surpluses and limit the Community funds that subsidize the agricultural sector. In other words, even though the West Europeans seem to have found a compatible mix of the public and the private in agriculture, the public's role in the Community may be at its optimal (most bearable) level, despite public political concerns, such as security and "defending new democracies," that justify the enlargement of the Community and support of overproduced specialized commodities.

North America exhibits a greater variety in its mixture of public and private features. The United States and Canada both value an open international market. The United States does not prefer and often denies its mixture of public and private wealth and power, even though it regularly supports farm incomes through a variety of measures, including deficiency payments and reserve accumulation of specialized commodities at some taxpayer expense and ideological disdain for government interference in the economy. A high degree of interpenetration between private interest groups and the government exists in the North American agricultural sector. American agriculture is a prime example of a subgovernment—or "cozy triangle" relationship—between producer groups, an executive agency (the Farm Bureau), and the national legislature. In the United States and Canada, as agricultural and food policy issues increasingly overlap with broader economic issues, consumers have also come to play a bigger role in determining policy outcomes.

During the 1970s the American government made an effort to let international market conditions determine prices and "get out of the farming business." But this adjustment hurt inefficient farmers, who, because of declining payments tied to lower target prices, were saddled with wide margins between commodity prices and production costs. Canada and Mexico, on the other hand, are more likely to acknowledge and accept the existence of government interference. As Professors Dickens and Moore point out, Canada has made a greater effort to stabilize prices and narrow the income gap for farmers, while Mexico directly intervenes in production in order to promote self-sufficiency and industrial development. Government subsidization of consumer food prices and no farm price supports result in production shortages and a growing dependence on food imports.

On the surface, the public and private distinction is theoretically irrelevant to centrally planned economies in Eastern Europe and the Soviet Union. However, this is actually not the case. Balaam and Carey highlight the significance of the private arena in Eastern Europe and the Soviet Union, where subsidiary farming and private plots are officially and sometimes "publicly" supported, both out of necessity for their contribution to agricultural production and, in the case of Eastern Europe, where they serve as institutions that enhance Eastern Europe's efforts to realize self-sufficiency and widen its distance from the Soviet Union. The bond between the public and private (to reintroduce the Japanese analogy) is, of course, more tenuous. In other words, the private sector is not an equal partner with collectivized agriculture, but in most cases exists at the behest of the public—i. e. the government. Production results have varied from mixed to poor in some countries. Many socialist economies are capable of achieving self-sufficiency and even surplus production but have been unable to overcome a number of natural limitations and political barriers associated with socialized agriculture. Efforts to increase consumption of meat and grain products require import dependency, which has grown over the past decade.

Except for the Soviet Union, neither food production nor food acquisition has been a major problem in the developed nations, although "pockets of hunger" exist within all developed countries. The crux of the interventionist issue is the political and economic costs of structural adjustments in the agricultural sector, which come after production levels reach satisfactory plateaus, or when a percentage of agricultural labor represents a good balance with other sectors. In Japan, Western Europe, and the United States, the question remains to what extent farming is a cultural and even aesthetic good that the polity wishes to preserve, which justifies support measures for inefficient farmers. In the Soviet Union the issue is directly related to balancing regime control over agriculture against decentralized measures that could increase production but also threaten leadership authority.

Insofar as the Latin American approach to political economy, discussed by Ganzel, generally recognizes the interplay of public and private wealth and legitimization of interventionist government policies, a great similarity exists between the Latin American countries cited and Western Europe. Intervention in the name of capitalism and markets, and even social democracy, is generally as welcome in Latin America as it is in Europe. Governmental intervention is a tool of public policy predicated on the belief that the state "remains the primary decision maker responsible for the welfare of its citizens in both actual and ethical senses."[4] Thus, in Latin America the problematic aspect of public and private wealth and power tends to fade in significance, but it does not prevent decision makers from using market mechanisms like higher consumer prices to mold policies along liberal economic lines of comparative advantage (as a strategy) to promote wealth and welfare gains.

According to Professor Weinbaum, all states in the Middle East appear to be interested in agricultural sectors for the political purpose of enhancing control and authority. Decision makers have given agricultural development only secondary attention and have even allowed it to languish, which contributes to a rather unusual situation of a widely privatized agricultural sector. Some countries have tried to preserve the role of private agriculture and at the same time pursue interventionist (public) roles in the agricultural sector. Since the early 1970s, as a result of increasing population and national income pressures, policymakers have begun to reevaluate their agricultural sectors. On the other hand, decision makers and bureaucrats committed to managed agricultural sectors still ignore the "behavioral effects on farmers of price relationships in regards to crop production and the effects of prices differing internally and in world markets."[5]

The PRC is still trying to define the role of agriculture and an optimum public-private formula to feed the populace. Since the 1949 revolution, the public-private relationship has gone through cycles of redefinition resulting in a greater acceptance of the role of private economies. These shifts correspond with leadership attempts to gain regime support, promote industrial development, and make agriculture subservient to ideology. Production increases correspond with more market-oriented policies in the national economy, outward-looking trade strategies, and government centralization effects.

In many ways the PRC is similar to many Middle Eastern countries: what looks like decentralized agricultural policymaking in communes aimed at enhancing the political economic stake of producers, especially smaller farmers, in the production process is actually "window dressing"—that is, it results in continued central authority control over the process. Also, the conflict surrounding many of agriculture's problems, or at least its role as part of a larger development strategy in the Middle East and the PRC, is often fought out in bureaucratic arenas. In the Middle East, bureaucrats and planners have resisted sharing decision-making authority with local or regional authorities out of a historical propensity to distrust their peers and an unwillingness to give up any of their own power or diminish their status. In the PRC, on the other hand, bureaucrats have been more foward-looking, and, as a group that competed with Mao for power, they have promoted more open market trade strategies.

In the low income market economies of East and Southeast Asia, food strategies manifest a combination of public and private features. Agriculture and food strategies have vacillated between efforts to achieve self-sufficiency and, as in the case of Indonesia, almost complete dependency on food imports paid for by oil export earnings. The growing influence of middle-class and urban interest groups has had a greater impact on support for market strategies in Indonesia, Malaysia, and Thailand, while the public sector has been more predominant in Burma. In all these countries and in Africa, the purposes served by agricultural policies and food distribution strategies are related either to the personal goals of leaders or to nation-building efforts that may include the use of food to help or hinder the poor or ethnic groups. The exact ratio of public-private measures is likely to reflect broader development and security goals.

In Africa the public-private mix is heavily weighted on the side of public measures. As in the case of East and Southeast Asian low-income market economies, food policies reflect an urban bias and a propensity to incorporate a range of government-sponsored activities into strategies that seek to enhance food production. These efforts, however, have not overcome many of the natural barriers that condition production. As economies and population have grown, accumulating surpluses has become a problem. Government interference, by requiring licenses of traders or the use of marketing boards, deters agricultural output.

The Middle East, the PRC, Latin America, and much of Southeast Asia (as well as the Soviet Union and Eastern Europe) are similar in terms of what Professor Weinbaum notes are the expected gains to be made, by either technological quick-fixes or political solutions like land reform that are easily outdistanced by growing populations and per capita incomes. This necessitates either imports of new technologies and agri-inputs to increase yields or imports of foodstuffs to cover food shortages. It also results in greater dependency and vulnerability to international market conditions. Therefore, a more common characteristic of countries in developing regions these days is the realization of the role played by agriculture in the development of the entire economy. However, many national leaders remain torn between development strategies that emphasize agricultural development or even self-sufficiency and strategies that forgo self-sufficiency and emphasize the development of comparative-advantage industries. It is often the case, especially when comparative-advantage strategies are chosen, that they are subverted, not so much by domestic fortunes or conditions, but by external conditions, factors over which national leaders are likely to have little control.

This is perhaps the most distinguishing feature between developed and developing regions—the former have more means available to absorb and thwart the effects of external influence on the domestic economy. The developing nations cannot as easily absorb the shock of fluctuations in international markets and supply conditions. Their populations are more often overtaxed or will not accept the political, social, and cultural costs associated with dislocation. LDCs have fewer opportunities to exploit their comparative advantage, and often adopt strategies that insulate their economy from the rest of the world.

The interplay of public and private exists at the international level too. Seth Thompson's chapter on international organization demonstrates that the problem is not so much one of defining the relationship of public and private wealth and power, since what is discussed at the international level is public and world organization relationships to national governments and indirectly to private sectors. According to Thompson, the function of international organizations like the FAO is to assist the national public level with data collection and advice and more or less to leave the national public and private interplay to be defined within national boundaries. Of course private (nongovernmental) international organizations that are relevant to food policies do exist; CARE and the Red Cross are two. However, their limited resources and restricted focus on humanitarian and crisis projects necessarily limit their scope and ability as "global" problem-solvers.

FOREIGN POLICY ARENAS: TRADE AND SECURITY

Table 10.2 summarizes the regional characteristics of agricultural trade and trade policies discussed by the contributors.

Agriculture and food play an increasingly greater role in the foreign policy arenas of both developed and developing regions. Three clear perspectives

Table 10.2 Regional Trade Characteristics

	Agricultural Trade Character	Agricultural Trade/Purpose	Agricultural Trade/Means
North America			
United States	exporter	surplus relief and export market development, help trade deficit; increased politicization and attempts to use food as a weapon	protectionist, prefer open, export subsidies; no marketing board; USDA efforts; trade embargoes
Canada	exporter	focus on export market development, surplus relief, help trade deficit, reluctant to form grain cartel	same means available as United States plus a marketing board
Mexico	importer, some exports	self-sufficiency, export market development	protectionist, marketing board
Western Europe	importer, some exports	farm support, meet consumer demands, surplus relief and export market development	protectionist, export subsidization of surplus commodities, CAP
USSR & Eastern Europe	USSR: accept importer role, E. Europe: less acceptance of importer role	USSR: import to meet consumer demand; E. Europe imports, but prefer self-sufficiency	closed, marketing boards

Latin America	importer, export some commodities	develop comparative advantage enterprises, stimulate investment, cover food shortages	protectionist, marketing boards
East & Southeast Asia	importer, export some rice	exchange commodities for energy resources, export market development	protectionist, marketing board, Executive ministries involved
High-middle income markets	importer	stimulate investment, meet consumer demand	open, little protection
Low-income markets	importer	imports, stimulate comparative advantage investment, cover food shortages	marketing boards (except Thailand), protectionist— becoming more so
Centrally planned economies	importer (PRC)	modernize agriculture, cover food shortages, (PRC); others closed, import substitution stimulate local investment	(PRC) open experiment
Middle East	importer	cover food shortages, stimulate investment	open, marketing boards
Tropical Africa	low-level trade	export cash crops	protectionist, state marketing boards

219

about trade's relationship to agriculture are evident in this volume: two are related to its economic dimension, and the third to its contribution to power.

Richard Ganzel and Lynn Scarlett share the view that trade is an "engine of growth" and that open market strategies enhance agricultural production and provide a number of other benefits to the countries they survey. Likewise they are both in agreement that self-sufficiency should not be the goal of countries whose comparative advantage would be the production of goods or development of industries in which they are more richly endowed.

Ganzel's analysis of Latin America links different trade strategies with agricultural production. Production decreased when industrialization goals were supported by import-substitution policies in Chile or by government manipulation of trade in Argentina. Ganzel's primary concern is with the side effects of market intervention on the efficient use and development of resources. Chile and Argentina have recently adopted trade policies that suggest a move back toward realization of their comparative advantage. Brazil's blend of "state capitalism with domestic and foreign participation" via more open market strategies has resulted in an increase in domestic and foreign investment in agriculture and "the very rapid growth of industry and agriculture."[6]

Lynn Scarlett notes that Africa has traded little with the rest of the world and that, more often, self-sufficiency by subsistence farming has been the norm. She believes that more open market trade strategies would serve to attract a source of credit the African nations need and would contribute to the infrastructural development of transportation industries and information services, help develop internal markets, stimulate domestic demand, and regularize the flow of supplies to internal markets.

A second and contrasting perspective about trade is developed by Professors Carey, Dickens, and Moore, who suggest that production success in their region is not due to open market strategies but, instead, is related to "the very closedness of these systems."[7] They all agree that agriculture defies rationalization, especially by employment of open market trade strategies. Furthermore, Dickens and Moore are of the opinion that free trade strategies are theoretical constructs not necessarily based on actual political-economic conditions that combine to determine trade policy. Free trade is always distorted by "domestic political parties, opinion, and interest groups," international market uncertainties, and "cycles of protectionism." Michael Carey adds weather vagaries and "human influences" to the list of diluting factors.

Dickens and Moore see the United States, Canada, and Mexico moving slowly toward a position that only resembles rationalization, or a tentative acceptance of free trade strategies, ready to draw back to a more protectionist position when domestic pressures demand it. Trade liberalization or open market strategies are tenuous positions that defy institutionalization, and Michael Carey's discussion of the CAP of the EC demonstrates his contention that many national interests are constantly in a state of redefinition. When combined with the rather unstable conditions of market supply and demand, policy acceptance by the entire Community is quite remarkable, but it is also likely to require reconsideration or readjustment in a short period of time.

Balaam's assessment of Asian economies and trade policies reveals instances of both the previous perspectives. Japan's trade policies correspond with both perspectives in that agriculture was rationalized, in order to promote industrial development in the early period. As in the case of other advanced industrialized economies, Japan's trade policies have recently been politicized by a number of conditions, including conflicting domestic interest group demands. Consumers want to hold down food prices, while producers vie for and receive higher food prices in political arenas where their disproportionate influence on price-setting policy has been institutionalized. Japan's agricultural trade policy also reflects an attempt to protect surplus products, which is justified on the basis of their contribution to self-sufficiency, national food security, and use as exchange for many of the raw materials and energy resources with which Japan is not endowed.

The high-income market economies of Asia have, to a great extent, adopted the type of trade strategies recommended by Ganzel and Scarlett. More open market trade strategies complement efforts to develop comparative advantage and commercial sectors and semimanufactured goods industries. Low consumer food costs also complement this strategy, because increased savings can contribute to investment in industry—in effect this strategy accepts becoming a food importer. The smaller centrally planned economies have adopted the import-substitution type of trade policies in an effort to enhance the growth of "infant industries." Because of the great amount of instability and war in Southeast Asia it is still too early to determine whether these policies have resulted in production gains or losses.

Conditions in the PRC support the Ganzel-Scarlett contention that certain types of trade policies ultimately contribute to inefficient and decreased production, as agricultural production dropped off drastically during the Great Leap Forward and Cultural Revolution campaigns. When more open market trade strategies were adopted, Western technology helped develop and improve agri-inputs and increased output yields. The PRC is now counting on more open market trade strategies to increase agricultural production and enhance the modernization of agriculture and industry.

The trade situation of the low-income market economies of East and Southeast Asia resembles that of the Latin American countries, which abandoned their agricultural comparative advantage and adopted industrialization strategies. As Balaam characterizes the situation, in many of these nations some of the points may by Carey, Dickens, and Moore are demonstrated, as it was a combination of political, economic, and social factors that led the adoption of industrialization goals. Leadership attitudes and values, natural limitations (particularly adverse weather conditions), tremendous population growth rates after World War II, bureaucratic structures and values left over from the colonial period, the emergence of a middle class in industrial and commercial enclaves, and the fascination with modernization and industrialization led to periodic shifts in strategies and a lack of attention to agriculture. In some of these countries, industrialization at the expense of agricultural development has since been judged an inappropriate strategy. Consumer prices have been raised to enhance special-ization and achieve agricultural self-sufficiency.

Richard Ganzel's work on Latin America is a compelling argument in favor of economic growth and the adoption of trade strategies that favor specialization and comparative advantage over self-sufficiency. However, the extent to which policymakers in developing regions can economically or politically afford to take Ganzel's advice and analytically separate "questions of economic efficiency . . . from those addressing cultural norms or the distribution of rewards within societies"[8] is questionable.

A focus on agriculture and trade in different regions of the world does not elicit any *definite* relationship between trade and hunger. For every situation in which open market trade strategies and complementary domestic food policies resulted in improved consumption, there are instances in which they did not (in terms of solving major food problems). Of course, it is not the strategy per se that results in increased import dependency, but a number of other domestic and foreign political economic (and even social) conditions that prevent nations from realizing the full benefits of an open market trade strategy. The more-developed economies like Japan, the United States, and Canada successfully adopted modified open market trade strategies to their physical, social, cultural, political, and economic environments. In poorer nations, these same environments may act as barriers to achieving higher growth rates or a more equitable and efficient distribution of food and income—hence the need for special strategies.

The third perspective about agricultural trade policies is the growing politicization of agricultural trade caused by the purposes agriculture and food serve in a rapidly changing international environment where the goal of self-sufficiency is emphasized more often due to security interests.

On the other hand, as discussed by Dickens and Moore, agriculture's role as a balancer of trade deficits in the United States and Canada is quite different from the role played by agricultural trade in the 1950s and 1960s. Until the mid-1970s American agricultural surpluses that made up American food aid to Western Europe and a number of Asian countries and that were later used as part of P. L. 480 served a number of much less politicized purposes—namely, as *assistance* to nations whose strategic position and regime viability was of great concern to the United States in the Cold War environment. As symbolized by the title of P. L. 480, "Agricultural Trade Development and Assistance Act," food was sold to develop markets and complemented producers' desires for surplus relief.[9] During the 1960s and early 1970s the use of food was part of humanitarian efforts to aid LDCs faced with overwhelming food production problems and shortages. It was generally felt that food aid would supplement their economies, relieve their governments of a number of immediate problems, and "get them over the hump" so they could attend to more long-range nation-building activities that would result in their development.

Due to a lack of success of these efforts and a number of other dramatic international events, including production shortages in the largest surplus-producing countries in 1972 and the 1973 OPEC oil embargo and price increase, agricultural trade and large yearly surpluses took on the role they have today: one in which export sales are used to balance trade deficits. For example, one out of every four acres of grain produced in the United States is marketed for export. This role is compatible with a new international environment with frequent but irregular commodity production shortages over the whole world, increased demand for food in LDCs, and a relaxation in East-West tensions. However, consumer concerns about adequate food reserves in surplus-producing countries have increased and were as much a part of the 1975 Moscow agreement that halted the Soviet "grain robbery" and the 1975 embargo on soybean exports to Japan as these measures were attempts to establish export markets or stabilize market prices.

Self-sufficiency and food security, then, have been added to a number of other justifications of farm support and protectionist trade policies in Canada, the United States, Western Europe, and Japan and have greatly politicized food trade policies and strategies in domestic political arenas that reflect increasing interdependencies among the Western market economies. This explains why agricultural trade protectionism has been such a difficult issue to deal with in such forums as GATT and UNCTAD.

The issues of self-sufficiency and food security have often risen to the top of the public policy agendas of many developing nations in Southeast Asia, Africa, and the Middle East. Some of these countries have been helpless in the face of shifts in the international distribution of wealth and power. Some Eastern European countries are now concerned about their vulnerability to international market conditions and have adopted policies in pursuit of self-sufficiency. But the Soviet Union has accepted its relatively new role as grain importer, which leads many to think that it is vulnerable to attempts to link its dependency on grain imports to demands for change in its domestic and foreign policies.

The growing vulnerability of nations and regions to international market price fluctuations and commodity shortages seems to provide food surplus producers, like the United States, Canada, Australia, Argentina, and France, with a distinct advantage in terms of insulating themselves from a number of political economic shocks associated with food supply problems and to provide them with a new type of food power. This power is due to food's potential as an instrument of foreign policy and is often determined by the degree of import dependency. Recently, there has been a good deal of discussion by public officials and academics about the use of food as a commodity weapon to counter OPEC influence, for example, or in a more positive fashion by distributing reserves in a responsible fashion, i.e., as part of an effort to establish some kind of new world order.[10]

Balaam and Carey's chapter on the Soviet Union and Eastern Europe and the discussion above demonstrates that the use of food as a "tool of foreign policy" did not begin with the current American grain embargo on the Soviet Union. Until 1963 food was used in support of the Cold War American embargo efforts on the East. After that time, increased East-West agricultural trade served that broader purpose of improving East-West relations.

The 1972 Soviet wheat deal was also part of a conscious effort to enhance Soviet-American détente and complemented a number of other domestic political economic objectives.

However, the political economics of the current American grain embargo against the Soviet Union support those who have previously argued that food would prove to be an inadequate *weapon* in the American arsenal.[11] A number of conditions in the current situation seem to support this argument.

First, the United States has not been able to control the actions of other producers[12] or grain shipments, 90 percent of which are monitored by the United States. The Soviets have found a number of other trading sources, including Argentina and Brazil, who are willing to "break the strike" and sell more expensive grain to the Soviets. Some grain may also have reached the Soviet Union through Eastern Europe or third countries. Second, it is questionable to what extent the embargo has created serious havoc for Soviet decision makers, i.e., forced them to reconsider their goal of increased livestock production and meat consumption. There are a number of alternative measures the Soviets could use to meet the grain crunch.[13] The United States must also consider the possibility that the Soviet decision makers would simply fall back on the old policy of forcing consumers to bear the greatest burden of the embargo by rationing grain and meat products even more than they are now.

Aside from the costs of the embargo to American taxpayers and farmers,[14] the embargo could prove more costly to the United States and ultimately to LDC food importers, who depend on low-priced grain to cover food shortages. An American famine reserve program, although well intentioned, could cause the international market to signal large carry-overs of supplies and result in land diversion programs, precisely at a time when world food production and consumption patterns in many regions have deteriorated significantly! If these reserves were released for anything but strictly emergency situations, they would amount to dumping food on the market and could deter local investment in food production projects.

Still greater costs the United States might incur are associated with changes in world food production and trade patterns and the United States' political relationship to its allies. Like the soybean embargo on Japan, the grain embargo may undermine dependence on American exports, stimulate production of those commodities in other countries, or drive importers to self-sufficiency, thereby weakening the position of the dollar on the world money market and losing a guaranteed market for American farmers. Likewise, the embargo may backfire on the United States if efforts to get other producers to cooperate with it require political concessions. Argentina's demands for changes in American human rights policies is a case in point. Also, given the trend of Soviet leadership attempts to respond to consumer demands via more decentralized agricultural policies, the embargo could ultimately justify recentralization and a return to self-sufficiency strategies. Thus, the grain embargo is a "one-shot deal" that depends on Soviet willingness to gamble against this year's estimated crop output, and could prove counterproductive to long-range American foreign policy goals and domestic welfare.

This discussion is not necessarily meant to indict the American grain embargo, and certainly not to imply acceptance of Soviet behavior in Afghanistan. Instead, it serves to delineate more clearly the boundaries of using food in an overtly political fashion that is, as a weapon. On the basis of what the authors have said and the discussion above, a number of points about the utility of food as a means to accomplish particular ends have been demonstrated.

First of all, there is definitely such a thing as food power. Nevertheless the extent to which food alone is a factor that can condition outcomes favorable to the West or the United States is debatable. Food is one political tool among many that, as part of a broader strategy, is most effectively used in a diplomatic or political payoff fashion. This is not to say that some day conditions could not be arranged in such a way that food would prove to be an effective weapon. But to extend the role of food—from an instrument of foreign policy to a weapon that in and of itself can change behavior—may not be possible and may have greater consequences for those who would attempt to use it than for those whom it is intended to hurt.

Thus, the second point is that the attempted use of food as a weapon has repercussions for users and recipients in terms of political and economic costs and benefits. To use food as an instrument of diplomacy necessitates the reconciling of a host of conflicting demands made on policymakers related to the purpose it should serve and to national, regional, and even broader interests. Finally, in this age of the balance of terror *and* commodity shortages, the propensity to use nonmilitary commodities as diplomatic instruments will persist. Thus, any calculation of power in international relations must include the availability of food to actors in an environment in which a hierarchy of vulnerability perpetuates a hierarchy of diplomatic influence. Such calculation of power also points to the greater role food plays in global interdependencies and especially the complex interdependence, that is, the problem of cost sharing and absorption, that exists among the industrialized developed nations.[15]

REGIONS INSTITUTIONALIZED

This section is an assessment of the institutional side of regional food politics. Highlighted are the forces that contribute to the institutionalization (and breakdown) of structures, as well as the effect they have on agricultural production and food distribution processes within specific regions.

The EC of Western Europe, COMECON of the Soviet Union and Eastern Europe, and ASEAN in Southeast Asia are the most institutionalized regional integrative efforts in existence. The main reason for the support of such efforts has been either security concerns, which applies to the EC and COMECON, or economic specialization, which applies to the EC and ASEAN. The EC and ASEAN have moved toward closer cooperation but must deal with different types of food problems, while COMECON has become frozen or has even disintegrated in proportion to the difficulty the Soviet Union has maintaining hegemonic control over its satellites.

Michael Carey highlights a number of security, ideological, hegemony, trade, and resource interdependence barriers overcome by EC members so that new "rules of the game" for agriculture could be established within a regional context of the CAP. Even though the EC is the most institutionalized and integrated regional organization, integration has its limits. Enhanced regionalism will be based on the benefits of further enlargement to individual members. At the same time, public considerations, especially security concerns and ideological preferences as well as opportunities for expanded internal regional trade, predispose the Community toward enlargement from nine to twelve members. There are, however, strains in the internal logic of protection of private welfare because CAP policy has reached its budgetary limits. Growing urban and industrial sector resistance to increased expenditures on the CAP is a sign of this strain. However, according to Professor Carey and given the particular agricultural sectors of the possible new members, who will not be important producers of northern European products (such as milk and meat) and are net importers of some northern produce, "their accession may relieve the pressure on some of the commodities which are either in surplus or bordering on self-sufficiency."[16]

The biggest advantage achieved by Western European agricultural regionalism is its contribution to the wealth of the region, which in turn is responsible for its being a major food importer. This importer status augments the EC's basic power as a unified region in international forums where food trade is discussed. Enlargement, which will add another 50 million people (many of them farmers) to the Community, could further strengthen the Community's hand in defending its internal policies, its trade practices, and most probably its power to recommend (at least for the rich producers) worldwide arrangements such as a wheat cartel, or alternatively an international grain bank.

The institutionalization of regional integration of the Soviet Union and Eastern Europe's economies seems to have reached its zenith. Members prefer to deal with one another as nonmembers on a bilateral basis. Many of the economic and political problems overcome by the EC have yet to be overcome by COMECON's members. The primary motivation behind Soviet integration efforts have dissipated due to its reluctance to share regional decision-making authority with its allies, who do not desire integration as much as they do autonomy. Instead, the bilateral relationship of each nation to the Soviet Union continues, which, on the basis of unfavorable terms of trade and political hegemony over Eastern Europe, allows the Soviets to exploit their neighbors. Other major obstacles remain, and regionalism has not resulted in a more efficient use of resources or specialization. For the Eastern Europeans, agriculture plays varying roles in each polity. For those with the comparative advantage, agricultural self-sufficiency has not been attained because ideological barriers related to attempts to adopt the Soviet "command economy" model of development have yet to be overcome.

On the other hand, linkages between ASEAN member nations have increased integration. Although a number of divisive issues remain, in particular intense nationalistic feelings, many differences have been overcome. There is a new breath of life in ASEAN, which has come to symbolize a new political future for its five members. Concern over the PRC has dissipated to some extent, and the PRC has made cooperative overtures to ASEAN. According to Balaam, high on the policy agenda are "tariff reductions, preferential trade agreements, . . . [and] . . . a surplus rice stockpile and distribution plan."[17] Another ASEAN goal is the promotion of political stability, which it hopes will provide the opportunity for national leaders to reform their economies and promote a better balanced development, one that includes a more important role for agriculture. Thus, domestic structural conditions and security concerns compel national leaders to adopt a mixture of self-reliance development and outward-looking trade strategies.

Although there is no North American common market, Dickens and Moore argue that a number of conditions exist among the United States, Canada, and Mexico that enhance a common regional perspective. There already exists a trend toward the development of cross-border trade and investment relations. In the interest of "regional" harmony, officials have often ignored the complaints of producers who compete with neighboring producers. Governments have also prodded their producers to conform to the marketing standards in neighboring countries so as to guarantee continued imports of these products.

Yet the United States and Canada remain competitors for grain markets. Bilateral relationships with grain importers still provide more benefit to each producer than do multilateral agreements. Problems remain for competition between America and Mexican labor and easing labor flow restrictions. Mexico also has its oil to use in carrot-and-stick fashion to extract economic concessions from the United States and Canada. But neither northern neighbor has adjusted to Mexico's oil power. Thus, the three North American polities are not in a position that would make integration mutually beneficial to all partners. Nationalism, in the case of all three partners, and different levels of economic development in the case of Mexico are powerful disincentives to further regional integration. These need to be overcome before formal institutional structures are created. Informal regionalism may intensify.

Three conclusions about regional organizations have evolved from this discussion. First, the food policies and problems of regional organizations are manifestations of aggregated national food problems contained within the region. If balancing public with private welfare is a predominant problem for agriculture in Western democracies, it is one the EC's CAP cannot easily solve. Community protection of producers results in surplus production and accumulation of specialty commodities. Thus, CAP policies result in costs and provide benefits for different Community members. As Professor Carey notes, this makes attitudes about integration often change in roller-coaster fashion. The disintegration of COMECON reflects the lack of a strong political underpinning for integrative efforts, except related to the political hegemony of the Soviet Union. As long as this situation persists, we are likely to see Eastern Europe reform the Soviet model in their nation-state to fit local conditions and to prefer bilateral rather than regional political and economic relations with other members. ASEAN must overcome obstacles created by nationalism, economic development variation, and severe natural limitations, and also a variety of development strategies governments pursue with too little planning.

Second, regional organizations may also have economic clout (as demonstrated in the case of the EC), but they cannot insure political stability in areas like Southeast Asia where conflict has been the norm of regional politics. Such organizations do not *produce* cooperation in itself, but *require* cooperation as a basis upon which to exist. Like the proposed Canadian-American wheat cartel, formal North American regional integration efforts have not gotten off the ground, which is due to the lack of a firm political foundation and lack of an outside security threat that would compel breadbasket countries to integrate.

Finally, regional organizations may help relieve hunger problems, establish food policies, and help development objectives such as economic efficiency and specialization. But they are not the only solution to such problems and may in fact compound them. LDCs who count on integration, efficient production, and intraregional trade development to solve their immediate problems will be disappointed. ASEAN could eventually play a major role in the promotion of the production of rice, in which the region has a comparative advantage. But it remains to be seen whether national leaders will continue to believe that industrial development alone, as is now stressed, without conscious efforts to enhance local food production and better distribution in Asia, will be able to overcome the effects of high population growth rates and crop fluctuations. Given the variety of Asian food systems, then, a regional food policy based on the growth-stage model could skew incomes and cause even more hunger. ASEAN officials are beginning to realize that industrial development cannot easily coexist with hunger and can even deter further economic development.

Policy Recommendations

One purpose of this volume is to use the regional approach to food politics as a foundation upon which to base policy recommendations that help relieve hunger problems. The regional approach points to the difficulty of reconciling natural limitations to production and distribution with both national and regional goals and capabilities, and thus to the difficulty of making suggestions that will always be "politically accepted courses of action for policy-makers." Yet a statement of problems and possible solutions is necessarily the first step toward actualization of policies that contribute to hunger relief.

Starting with food problems in specific locales and moving up through national and regional levels, it is easier to target recommendations to particular problems and focus on the major trade-offs and repercussions decision makers face when considering policy choices.

The greatest numbers of the world's hungry people are located in the U.N.-designated Most Seriously Affected (MSA) nations in Asia, Oceania, and Africa. There is also a good deal of hunger elsewhere in the world, in so-called pockets of hunger. Drawing on the chapters above, attempts to deal with hunger range from intentional starvation in Cambodia, to an acceptance of some amount of hunger in many developed nations which prefer to use market forces and not structural programs to manage hunger. In some countries where the hungry are isolated from the rest of the population, there is no political rationale for attending to agriculture or for designing more coherent food distribution strategies. Most often, hunger is a secondary concern of government leaders, especially in developing nations which either exploit agriculture and extract its production to invest in industrial development or fail to reform agriculture in order to maintain regime support based on land tenure systems.

The first policy issue, then, is to place hunger relief on the policy agenda and to make it a primary goal in nations with large numbers of underfed people. Some contributors to this volume point to a significant change in the attitude of many national leaders in all types of polities, especially in LDCs, who realize that agriculture plays an important part in achieving economic growth, industrialization, and national security. Aside from national security, which has not always been a basis upon which to formulate agricultural policies, the goal of economic growth and industrialization should cause reexamination of agriculture's role in the economy. It is evident that poverty and development can exist side by side, but it may be wasting human resources and deterring economic growth to allow this situation to continue. For real growth to occur, the 40 percent of the population below the poverty line must be brought into the economy through distribution measures.[19] Likewise, the rural sector cannot be viewed as a marginal sector of the economy, or can farmers be allowed to remain outside the mainstream of national politics.

There is hardly a developing region where the forces of modernization have not brought about a compulsion for people to expect and consume more. The problem is one of adopting an appropriate economic growth and development strategy, one that incorporates agriculture and alleviates hunger and poverty. Still, it is evident that the overwhelming hunger and malnutrition problems of the poorer nations may not be solved by implementing a traditional development strategy that relies on market forces to solve food problems via "trickle down" effects a strategy in modified form that has been more successfully implemented in the developed regions of the world or by attempting to adopt an egalitarian distribution strategy.[20]

Market-oriented strategies without supplemental distribution efforts are inappropriate for poorer countries because they lead to dual economies, skewed incomes, and underconsumption, and worsen the chances for balanced growth. Higher food prices do not always help curb hunger and malnutrition; instead, they often leave nations dependent on food aid programs and/or on barely existent exchange earnings to fill both investment gaps and the immediate needs of those who are the worst off. An almost exclusive focus on growth to the exclusion of distribution will not create a much broader type of sociopolitical development. Latin America, for example, may have become reliant on comparative advantage strategies without giving enough attention to distribution, against a background of mass poverty and much malnutrition and hunger that may not be overcome by specialization strategies alone.

The case of China[21] demonstrates that countries cannot foresake growth for ideological purposes. A nation cannot necessarily overcome the natural and political barriers that constrain agricultural production and food distribution by adopting a model of development that emphasizes popular participation, a communal production system, and an Agriculture First strategy. Eastern Europe and the PRC seem to have learned this lesson. Specialization and efficiency are valuable in terms of their short-term growth-producing features as well as for what Ganzel and Scarlett call investment attractions and infrastructural development features. Thus, specialization, economic efficiency, and comparative advantage have a place in development strategies, but not to the exclusion of distribution programs.

Recommendation 1: Strategies of self-reliance (as distinguished from self-sufficiency) contribute the most to short- and long-term hunger relief and economic development. Economic growth and distribution strategies need not be seen in a zero-sum fashion, or as necessarily contradictory. The best development strategy for poor countries is the "self-reliance" strategy from Asia (discussed by Balaam) because it attempts to account for idiosyncratic domestic sociopolitical-economic conditions.[22] Policymakers in LDCs should focus on distribution as much as on growth. The issue is not economic efficiency as much as it is balancing efficiency with the possible greater costs of side-effects related to unbalanced development. The trend of agricultural exploitation should be reversed and, where possible, industry should serve agricultrue in much the same way as the PRC is gambling that it will. Two positive results of such a strategy are its economic effect on targeted groups, which stands to help chances for balanced economic growth, and the improved income distribution and hunger relief that could ultimately decrease the need for closed trade measures, such as import-substitution policies that are often used to insulate an economy.

Weinbaum and Balaam point out however, that policies adopted in accordance with any new model of development are likely to disrupt the distribution of wealth and power within developing nations because of the degree of societal and institutional transformation they require. On that basis alone, they are likely to be opposed. They are also likely to be stonewalled by bureaucrats who resist change or the sharing of responsibilities with regional or local officials. But elite interest, in particular, and their political support of the government, must be balanced with the need for increased agricultural production and better distribution.

Recommendation 2: Because economic development cannot be separated from political development, farmers and peasants should be integrated into the policymaking process at all government levels. In countries where self-reliance can be achieved, the integration of farmers into the policymaking process will probably cause them to be a more effective and satisfied economic group. The result may very well be more stable (if not spectacular) political and economic development.

Another issue is the role governments should play in relation to the type of development strategy chosen. Lynn Scarlett recommends that governments interfere less in the agriculture production and food distribution process. She makes a good case for an arrangement of policies that might better enhance production. Although the capitalist model seems to complement a decentralized political system, her chapter indicates there is no necessary correlation between agriculture and government organizational structure. Regardless of the mix of public and private policies, poor countries such as those in Africa, would benefit from some governmental policies to facilitate a smooth working, decentralized, production and distribution system that Scarlett favors.

The balance of public and private wealth and welfare that complements local conditions is most important. Government can facilitate efficient production by assessment of the problems that deter production or prevent the realization of comparative advantage. National governments can adopt policies that contribute to an improved farm system, the education of farmers, an increased supply of agri-inputs to farmers, and access to domestic and foreign markets.[23] Government can prevent many of the bottlenecks that occur between agriculture and other sectors so as to realize the benefits of sectoral interdependence, or can manage the national "rules of the game" by setting the pace of the intersectoral exchange process so that private and public welfare are equally accounted for.

An example of positive centralization is land reform in the PRC, where it served a political purpose and was also part of a total transformation of society that was initiated and implemented by a strong centralized government. Even if food production in the PRC has not been as great a success as leaders hoped, collectivized agriculture has been a good vehicle to insure that benefits are actually shared by most of the peasantry. Thus, some amount of centralization at the highest levels is necessary to initiate and support the types of policy reforms necessary to increase agriculture production and enhance food distribution efforts. It is not socialized agriculture per se that deters production. In many cases socialism can enhance output by adding labor to the production and distribution processes. The case of Hungary demonstrates that what is most important is the *flexibility* of any model—that it can be reformed to fit local production conditons.

But there may come a time when overcentralized socialist agricultural systems are either "worn out" or determined to be no longer appropriate to local conditions. Overcentralization can deter production and, in and of itself, create bottlenecks, as has been the case in the Soviet Union and on occasion in parts of Eastern Europe, the Middle East, Asia, and Africa. Therefore, decentralization as close to local producer levels as possible should be a goal of governments concerned with improving agricultural yields and production efficiency. Government's role should be to facilitate agriculture with back-up systems. Officials in the Middle East, the low-income market economies of Asia, and Africa have created policies for credit, cooperative loan societies, and agricultural development banks. Still, some governments have not followed through with other measures to help the greatest number of farmers receive these benefits, which is due to the cross-cutting purposes these policies serve.

Most LDC governments, because they are so tightly bound up in the nation-building process, cannot afford to let market forces dominate the process. Thus, instead of less centralization, we are likely to see more selective centralization as governments are compelled by internal or external maladies to accept more responsibilities in the narrower areas mentioned above. Centralization corresponds with development and modernization, and it may be that relative decentralization is a luxury only the more developed nations can politically and economically afford.

Recommendation 3: Centralization—the integration of public and private production and distribution units and/or selective intervention by government—should serve not only agricultural development but also comparative advantage and specialization in order to generate income related to natural resource capabilities. But specialized industries should not be promoted to the exclusion of basic wealth and welfare programs. As domestic and foreign arenas continue to tighten, it becomes clear that in some cases external conditions influence and even determine the success of development strategies. Most developing countries are dependent on other nations for trade, aid, and investment or, in a neo-imperial manner, are politically subordinate to a hegemonic power. Trade strategies vary considerably. Given our preference for a mixed development strategy of self-reliance, open market trade strategies are also preferable. But protection-ism for the sake of infant industry development will continue.

In some developing nations, more outward-looking open market trade strategies would enhance agricultural production if concentrated on a national comparative advantage. Yet the success of the high- to middle-income market Asian economies that have significantly increased their economic growth (and whose populations now eat more and better quality food as a result) is predicated on the absence of structural and political bottlenecks that could prevent economic growth and adequate food distribu-tion. In many cases their small territorial size enhances easy access to internal markets, while their trade success is conditioned by "preferential access to large markets in the developed world."

The Soviet Union and Eastern Europe resemble the more healthy developing nations which are at the crossroads of choosing strategies to realize their goals, as recent trade policies reflect consumer influence and leadership attempts to meet consumer demands. Most Eastern European trade policies are more open than those of the Soviet Union in order to attract Western technology, consumer goods, and foodstuffs, and to increase agricultural output because they cannot rely on Soviet grain and other agricultural exports. In these countries we are likely to see greater efforts to achieve self-sufficiency, either on the basis of comparative advantage or because of the necessity of trimming deficits that soak up potential specialty investments. Some Eastern European nations have recently increased production via a number of measures that include a rise in consumer prices.

In contrast, the poorer underdeveloped nations have large populations, high birthrates, and undeveloped markets or poor distribution programs.[24] Mass poverty and even starvation exist in much larger enclaves. These nations also lack the centralization of administrative government functions necessary to bring about fundamental change. In low-income market Asian economies, in the Middle East, and in Africa, the problem remains one of reconciling natural endowments that would suggest specialization of certain products, with the many concerns, including national security, that compel them to seek self-sufficiency. The development of MSAs is dependent not only on their relationship to other nations, but may place them at the mercy of other nations. For these reasons, many LDCs and MSAs will more than likely adopt protectionist trade policies. They cannot wait for a "trickle down" that may never come due to the low earnings suffered as a result of worsening terms of trade for their products and declining amounts of aid. Ultimately, it is the advanced, industrialized, developed nations that largely influence the terms of trade for LDC exports, aid, and investment activities in LDCs.

Recommendation 4: The developed nations should offer better terms of trade for poorer nations whose development is a foundation for the continuing economic vitality of rich nations. The Lomé Agreement is a good model of the type of agreement the rich should be willing to make with LDCs. A major issue related to trade is how to reconcile the protectionist trade policies[25] of the developed nations with development objectives of LDCs that would benefit more from open market strategies. The issue hinges on how much political and economic dislocation the developed nations are willing to accept. Heretofore they have been unwilling to absorb external economic dislocations and have exported them back into the international economy via the use of any number of protectionist measures. The industrialized nations, especially the Western democracies which are both food exporters and importers, have realized relatively efficient development of agricultural comparative advantages. But inefficient industries in these more pluralistic polities are likely to bear the greatest cost of dislocation. These industries and related interest groups try to rebalance the public-private welfare ratio in decentralized policymaking arenas to gain protection that insulates them from foreign competition or fluctuations in international market conditions. Moreover, due to increasing interdependencies between developed Northern countries, free trade is not likely to become any more of a reality than it already is.

It is in the economic interest of developed nations to adopt measures of government interference of another sort: export subsidies to enhance market development. Dickens and Moore note that American grain reserves might serve the purpose of "world development." Areas with hunger and poverty could be cultivated by food exporters as markets for overproduced goods. If exports were sold to nations like the Soviet Union which have the resources to pay higher prices, funds could be diverted to poor nations in the form of concessional sales of needed products or outright grants that would bolster their economy. Poorer nations could better realize their comparative advantage, even if labor were their only specialization.

Likewise, foreign aid, especially food aid, can play a major role in world development. Not only does food aid assist producers with surplus relief but it also stands to help poorer nations temporarily in emergency situations.

Recommendation 5: Food surplus producers should agree to a multilateral grain reserve, funded by surplus producers and food importers, to be used only in emergency relief situations. Food assistance over a longer period of time is reconcilable with local development efforts as long as it is distributed to those who need it most, thereby substituting for subsidization programs and allowing investment monies to be channeled into comparative advantage enterprises. In most situations, food aid would benefit from closer cooperation between national and international organization distribution units.

Recommendation 6: For humanitarian reasons, the rich nations should increase their levels of bilateral and multilateral foreign aid and food assistance to supplement LDC development projects. Yet protectionism and reductions in foreign aid are most likely to remain the order of the day. Therefore:

Recommendation 7: LDCs should not rely on aid to supplement or subsidize their development projects or strategies. Governments in poor countries should encourage increased productivity in rural and metropolitan areas where possible. Added to the difficulty of reconciling domestic policies with foreign policies in developed nations is the difficulty of separating economic and political or security concerns from trade and aid considerations. Although protectionism has led to overproduction, most grain goes to those who can afford it. The 1972 Soviet wheat deal demonstrates the extent to which grain sales were part of efforts to institutionalize East-West détente and redress the American balance-of-payments problem. The current American grain embargo has repercussions far beyond its immediate attempt to use food as a weapon. Given the balance of terror between the world's two military superpowers and the regularity of commodity production shortages, food is likely to be considered most often in terms of its contribution to the goal of national (food) security and as a tool of foreign policy. Food adds another dimension to the problem of complex interdependence.

Recommendation 8: Food surplus producers should not attempt to use food as a weapon because it works against the interest of a worldwide system of comparative advantage and specialization. Stated otherwise, because of food's political economic dimension, food policies are apt to be politicized even further in the near future. Reconciling national interests with a strategy that contributes to LDC development without excessive developed nation expense will be difficult to achieve. Decision makers and citizens in the developed nations may be well intentioned, but not because of a feeling of moral responsibility or obligation will they be willing to absorb the costs of dealing with food problems in other nations.[26] A concern for their *own* economic welfare and security is the primary reason for a significant change in the relationship of the rich to poor nations.

From the perspective of the developing nations, retrenchment toward protectionism is a political instrument that increased Northern domination over the South. At the same time, trade liberalization is viewed as an imperial scheme to dominate the international market structure, which cannot be separated from the international distribution of power. Two relatively new conditions could yet reconcile Northern and Southern interests. They are the earth's finite natural resources, and the growing interdependence among nations. Interdependence improves the position of Southern nations which are not hindered by the balance of terror and which possess many of the natural resources and raw materials that the developed nations (although they are not dependent on them) prefer to acquire instead of developing expensive alternatives. This situation may condition future food policies and provide a justification for dealing with hunger on both an economic and a political security basis.[27]

Just as interdependence compels nations to cooperate with one another, so it has a sensitivity dimension that promotes conflict. The chances are equally great that disagreement over concerns such as protectionism, multilateral commodity agreements, and a world grain reserve will continue. Bilateral agreements are preferred by nation-states over long-term, multilateral agreements, since short-run nationalism and state interests still pervade international politics. Regional institutions and organizations could enhance efficiency via specialization, but they cannot solve food problems related to ethnic and national rivalries, the variety of purposes agriculture serves in different economies, and economic dislocation problems the regional members face.

Recommendation 9: Regional organizations should promote interregional trade specialization and economic growth, if only to introduce some kind of planning to the development process. Regional development plans should also include hunger and poverty relief programs. Likewise, international organizations should continue the useful functions they serve but also should pressure national leaders forcefully to adopt hunger relief programs and strategies. Institutionalized integrated regionalism will succeed to the extent that specialization benefits are mutually shared and outweigh integration costs. Should a regional organization exist it must carefully consider appropriate food and agricultural policies as part of the regional development strategy and separated from other considerations.

It stands to reason that what Michael Carey calls establishing new rules of the game, at the level of global interdependent relations, will be one of the most pressing problems faced by all nations in the near future. While conflict cannot be eliminated, it will have to be better managed to account for the greater number of people whose wealth, welfare, and security concerns are being dealt with in larger political arenas. Finding solutions, or accepting the cost of solutions to hunger problems, is unlikely to dominate any global agenda *unless* agriculture and food problems are directly related to national wealth, power accumulation, and distribution concerns of the rich nations. Ongoing efforts to solve immediate problems are attempts to change, over a long period of time, a host of factors that contribute to hunger but that cannot be separated from broader political considerations.

Notes

1. A number of works have substantiated this argument quite well. One of the better statements in Susan George, *How the Other Half Dies: The Real Reason for World Hunger.* (Montclair, N.J.: Allanheld, Osmun & Co., 1977).

2. There are a multitude of such schemes. Some of the more popular solutions are birth control, the Green Revolution, and even triage. Most of these solutions are based on the assumption that those with massive hunger problems are responsible for the problem and that they should shoulder the greatest burden to solve the problem at little or no cost to the rich nations. Furthermore, this argument has recently been restated in terms of the "burden" some nations are to the international food system. See for example, Robert Paarlberg, "The Soviet Burden on the World Food System: Challenge and Response," *Food Policy* (November 1976). We feel that hungry people anywhere are not a burden but an integral political, economic, and social part of the international system.

3. See, for example, Jack Shepherd, *The Politics of Starvation* (Washington, D. C.: The Carnegie Endowment, 1975).

4. See Richard Ganzel's chapter on Latin America in this volume.

5. See Marvin Weinbaum's chapter on the Middle East in this volume.

6. Richard Ganzel, op. cit.

7. See Robert Dickens and Richard Moore's chapter on North America in this volume.

8. Richard Ganzel, op. cit.

9. There are a number of works on P. L. 480. One in particular that discusses many of its dimensions from a food policy perspective is Peter G. Brown and Henry Shue, *Food Policy: The Responsibility of the United States in the Life and Death Choices* (New York: The Free Press, 1977).

10. See for example Richard Falk, *A Study of Future Worlds* (New York: Free Press, 1975).

11. For a good discussion of the constraints posed on food producers who attempt to use food as a weapon see Robert Paalberg, "Food, Oil, and Coercive Resource Power," *International Security* 3, no. 2 (Fall 1978): 3–19.

12. Shortly after the embargo was announced, Western Europe, France, Australia, Canada, and Argentina announced that they would not replace the grain the United States would have exported to the Soviet Union, beyond that already committed to export as part of bilateral agreements. But the EC countries' trade in industrial goods is important enough for them to be reluctant to comply with American policy in agricultural trade. The Soviets could buy up EC surplus milk products, beef, and chickens if they were willing to pay higher EC prices on EC

butter and beef. Furthermore, despite a more recent embargo on phosphate exports used to produce Soviet fertilizer, the Soviets continue to import certain types of agricultural items from the United States that could be used to replace feed or meat.

13. At least four alternatives exist for the Soviets to meet the grain crunch. They could (1) speed up the slaughter of animals, which would temporarily improve meat supplies to Russian consumers but could bring about a meat shortage in a year or so; (2) place slaughtered animals in cold storage to smooth out the glut of meat; (3) purchase other types of animal feed (at higher cost), such as soybean meal from Brazil; and (4) cut into grain reserves, which is a risky step to take since the food security of the Soviet Union would be jeopardized.

14. American taxpayers have incurred the cost of government reserve accumulation programs, commodity subsidies, and farm loans to cover production costs to the tune of an estimated 3 billion dollars. Still other costs to the United States might be expected from export expansion and P. L. 480 programs that could soak up some of the surplus. Some of the grain normally purchased by the Soviet Union will undoubtedly go to importers like Mexico, which lost its supplier Argentina to the the Soviet Union, and to Japan, Italy, Spain, Portugal, the PRC, and a number of Third World countries. But these sales may not equal Soviet purchases, and it may only be the 1980 drought in the American Midwest that relieves the United States of its surplus problem!

15. For a more detailed discussion of the concept of complex interdependence and its significance in current international politics, see Robert Keohane and Joseph Nye, *Power and Interdependence*, (Boston, Mass.: Little, Brown and Company. 1977).

16. See Michael Carey's chapter on Western Europe in this volume.

17. See David Balaam's chapter on East and Southeast Asia in this volume.

18. See Michael Carey's Introdution to this volume.

19. For an excellent discussion of the relationship of mass poverty to development and the difficulty of adopting appropriate development strategies, see Mahbub ul Haq, *The Poverty Curtain: Choices for the Third World* (New York: Columbia University Press, 1976).

20. As Cheryl Christensen rightly notes, economic growth or even increased agricultural production alone does not guarantee adequate or even efficient distribution due to the structure of the market itself. See Cheryl Christensen, "World Hunger: A Structural Approach," *International Organization* 32, no. 3, (Summer 1978): 745–74.

21. See David Balam's discussion of the China model in the East and Southeast Asia chapter.

22. For a discussion of the self-reliance model see Frances Moore Lappé, *Food First, Beyond the Myth of Scarcity* (Boston: Houghton Mifflin Co., 1977). For a detailed discussion of specific features of the model see Sterling Wortman and Ralph Cummings, Jr., *To Feed This World: The Challenge and the Strategy*, (Baltimore, Md.: Johns Hopkins University Press, 1978), especially pp. 233–430. The problem with these studies is that they largely fail to link the self-reliance strategy to conflicting interests and strategies at higher government levels that must be reconciled for the strategy to work.

23. Wortman and Cummings, *To Feed This World.*

24. For an overview of conditions in LDCs, see the International Food Policy Research Institute, *Meeting Food Needs in the Developing World: The Location and Magnitude of the Task in the Next Decade*, Washington, D. C., Research Report no. 1, February 1976, and Wortman and Cummings, *To Feed This World.*

25. For a good discussion of agriculture trade protectionism, see FAO, "New Protectionism and Attempts at Liberalization in Agricultural Trade," *FAO Commodity Review and Outlook 1979–80*, no. 17, pp. 109–22.

26. This issue of moral obligation and hunger is debated in William Aiken and Hugh La Follette, eds., *World Hunger and Moral Obligation* (Englewood Cliffs, New Jersey: Prentice-Hall, 1977).

27. Haq, *The Poverty Curtain.*

Index

Adelman, Irma, 139n
Afghanistan: agrarian reform, 158, 159; agriculture, 144, 148; USSR invasion of, 62, 74, 225
AFL-CIO, 33
Africa (tropical), 166–88; agricultural policies, 177–86, 220, 223, 232, 233; cooperatives, 181–82; credit, 176–77, 180–81; export market, 172; farmers, characteristics of, 172–73; food crisis, 168–72; food market, 175–77; food problem, 167–86, 216; grading standards, 183–84; hunger gap, 169–70; hygiene, obligatory, 183–84; land ownership, 176–77, 180–81; large and small-scale farming, 173; licensing of traders, 180; marketing boards, 181; marketing problems, 173–74; minimum wages, 183; population growth, 169; price ceilings, 183; price controls, 179–80; prices of farm products, 176, 179; redistributive schemes, 170–71, 182; rural incomes, 171, 182, 183; transport problems, 174
Agricultural employment: in Eastern Europe, 68, 69, table, 70; in European Community, 47n
Agricultural policy (see Food policy)
Agricultural trade, 217, 200–25, table, 218–19
Aiken, William, 237n
Albania, 69
Algeria, 144, 148, 158, 163
Allen, David, 47n
Allende, Salvador, 93, 95–96
Anderson, Charles W., 85, 103n
Anderson, James E., 28n, 29n
Argentina, 91–92, 96–98, 102, 137, 220, 223, 224; food production, 89–90; income per capita, 89; land ownership, 96; political history, 96–98; USSR grain imports from, 59, 77, table, 61
ASEAN (see Association of Southeast Asian Nations)
Asia, East and Southeast, 106–42, 209 (see also China; Indonesia; Japan; other countries); agricultural trade policies, 221–22; centrally planned economics, 118–26, 131–32; food policies, 216, 217; food production, distribution, and prices, 107, 110–11, 114, 126–38, tables, 108, 114, 128, 223; gross domestic product distribution, 110, 136, table, 113; high- and middle-income market economies,

110–26, 130–31; land reform, 129, 134, 136; low-income market economies, 126–32, 232, 233; per capita dietary energy supplies, 110, table, 109; population and income growth, 110–12, 119, 127, table, 111; population structure, urban and industrial, 110, table, 112; regional institutions, private and public, 135; regional structure and food problems, 134–38; rice production and distribution, 136–37
Asian Development Bank, 135
Asian Industrial Development Council, 135
Askari, Hossein, 165n
Association of Southeast Asian Nations (ASEAN), 134–38, 208, 225–28
Augustini, Gunter, 140n
Austin, James E., 89, 103n, 205n, 206n
Australia, 137, 223; USSR grain imports from, 59, 62, table, 61
Autarky, 45–46n
Avery, William, 47n
Aziz, Sartaj, 206n

Balaam, David N., 7, 24, 29n, 47n, 208, 214, 221, 223, 227, 230, 237n
Barnett, A. Doak, 140n
Bauer, P. T., 182, 188n
Bellof, Nora, 80n
Benjamin, Gary L., 28n
Bennet, M. K., 187n
Berand, Mehmet Ali, 47n
Bhutto, Z. A., 155
Bill, James A., 165n
Black Sea, 144
Boerma, A. H., 195, 196, 199, 205n
Bohannan, Laura, 187n
Bohannan, P. J., 187n
Bolivia, 91
Bornstein, Morris, 78n, 79n
Brady, David W., 28n
Brannon, Russell H., 104n, 105n
Brazil, 77, 85, 91, 92, 99–102, 220, 224; agricultural development, 99–102; food production, 90; income per capita, 89; iron and steel industry, 100–102; Volta Redonda complex, 100
Brezhnev, Leonid, 49, 55–57, 64
Brittany, 31
Brown, Lester R., 14, 28n
Brown, Peter G., 28n, 29n, 236n

238

Brzeski, Andrzej, 79n
Buenos Aires, 96, 97
Bulgaria, 68, 69, 71, 73, 74
Bullock, Charles, 28n
Burma, 126, 127, 129, 132, 134, 216

Cambodia, 118, 131, 135, 229
Cameroon, 186
Campbell, Keith O., 187n, 188n
Camps, Miriam, 45n
Canada, 137; agricultural policy, 11–29, 214,
    220; Agricultural Stabilization Act, 15–17;
    agricultural trade, 222, 223; consumer
    movement, 18; exports to Japan, 24; farm
    income maintenance, 12, 18; farm product
    prices, 15–18; farm products, reserve policy,
    19–20; interdependence with US, 20–23; la-
    bor supply, 18–19; trade liberalization,
    23–24; trade with Poland, 71, 73; USSR
    grain imports from, 59, table, 61; Western
    Grain Stabilization Plan, 15–17; Wheat
    Board, 19–20, 25
CAP (see Common Agricultural Policy)
CARE, 217
Carey, Andrew Galbraith, 165n
Carey, Jane Clark, 165n
Carey, Michael J., 7, 29n, 46n, 47n, 214, 220,
    221, 223, 226, 228, 236, 237n
Carter, Jimmy, 62
Caspian Sea, 144
Chaudhry, M. G., 157, 165n
Chile, 85, 91–96, 102, 220; agriculture, 92–95;
    income per capita, 89; land ownership, 92;
    mining, 93–95; political system, 93–96
China, People's Republic, 107, 118–26,
    131–33, 137, 138, 209, 230; agricultural pol-
    icy, 110, 118–20, 122–26, 132, 216, 217,
    221; Cultural Revolution, 120–21, 124, 132,
    221; food imports, 121–22; food production,
    119–26, 132, figure, 121; foreign assistance in
    modernization, 137–38; Great Leap For-
    ward, 120, 123, 132, 221; hunger and malnu-
    trition, 120, 126; industrial development,
    122; land reform, 231; Machine Tractor Sta-
    tions (MTS), 123; modernization of agricul-
    ture, 124–25; political developments,
    122–26; population control, 119; relations
    with Mexico and Canada, 26; relations with
    USSR, 120, 126, 135
Christensen, Cheryl, 139n, 187n, 205n, 237n
Ciria, Alberto, 104n
Clark, M. Gardner, 77n
Clarke, Roger, 78n
Cohen, Benjamin J., 139n

Colombia, 89, 91
COMECON (see Council for Mutual Economic
    Assistance)
Committee of the Professional Agricultural
    Organizations in the EEC (Comité des
    Organisations Professionelle Agricoles de la
    CEE—COPA), p. 33
Commodity Credit Corporation, 19, 22, 25,
    62, 73
Common Agricultural Policy (CAP), 26, 27,
    31, 33–34, 37–45, 220, 226–28
Common Market, 214
Cooper, Richard N., 8
COPA, 33
Council for Mutual Economic Assistance
    (COMECON), 73–75, 208, 225–26, 228
Craig, Daniel, 165n
Crosson, Pierre R., 95, 104n
Cuba, 26
Cummings, John T., 165n
Cummings, Ralph, Jr., 237n
Curtis, Thomas B., 46n
Czechoslovakia, 59, 67, 68, 69, 71

Dalton, George, 187n
Deng Xiaoping, 124
Denmark, 92; ham exports, 46n
Destler, I. M., 29n, 205n
Diaz-Alegandro, Carlos F., 104n
Dickens, Robert, 6–7, 208, 214, 220, 221, 222,
    227, 234, 236n
Donnelly, Michael, 139n
Dorjahn, V. R., 187n
Drilon, J. D., 141n
Dumont, Rene, 182, 187n, 188n
Duncan, E. R., 186n

Eastern Europe, 6; agricultural employment,
    68, 69, table, 70; agricultural policy, 48–49,
    65–77, 214–15, 232–33; agricultural produc-
    tion, 65, 67, table, 66; credit from foreign
    nations, 73; currency, 74; labor force, 68, 69,
    table, 70; land ownership and collectiviza-
    tion, 67, 69–70; land reforms, 67–69; politi-
    cal reforms, 67–71; population and GNP
    growth rates, table, 56; regional integration,
    226–27; trade patterns and policies, 71,
    73–75; U.S. agricultural exports to, table,
    72; U.S. trade embargo against, 71, 73;
    USSR grain exports to, 59, 71, table, 58;
    USSR grain imports from, table, 61
Eberstadt, Nick, 140n
Echeverria Alvarez, Luis, 20
Eckstein, Alexander, 140n

Economic and Social Commission for Asia and the Pacific (ESCAP), 135
Economic and Social Council (ECOSOC), 192
Ecuador, 91
Edelman, Murray, 205n
Egypt, 164; agrarian reform, 158; agriculture, 144, 147–48; cooperatives, 156; government departments, 153; Open Door policies, 161; subsidy for basic foodstuffs, 199
Eicher, Carl, 188n
Eiland, Michael D., 141n
Elek, Peter, 79n
Enke, S., 187n
Enloe, Cynthia H., 141n, 142n
Esseks, John D., 88, 103n
Ethiopia, 174
Etzioni, Amitai, 8n
Europe: Eastern (see Eastern Europe); food policies, 30–47, 214, 223; interdependence in, 31–34; relations with China, 137; relations with Mediterranean area, 33, 34, 37, 40–45; relations with U.S., 33, 34, 37–40
European Community (EC), 208, 214, 225, 226, 228; agriculture, 35–38, 47n; COMECON and, 74; food imports, 35–36; food policy, 30–31, 34–45; grain sold to Eastern Europe, 71; subsidies in U.S. relations, 39–40
European Currency Union (ECU), 35
European Monetary Union (EMU), 34, 35
European Political Cooperation (EPC), 33–35, 38, 41, 42, 44, 45
Expanded Program of Technical Assistance, 191

Fagen, Richard R., 88
Fainsod, Merle, 78n
Falk, Richard, 236n
Fallenbuchl, Z. M., 80n
Feder, Ernest, 86, 87, 103n
Fender, Frank A., 104n, 105n
FEOGA, 37
Ferrer, Aldo, 104n
Field, John Osgood, 162, 165n
Field, Robert, 120, 140n
Fienup, Darrell F., 104n, 105n
Fischer, Lewis A., 22, 29n
Fisheries: Iberia, 43; Ireland, 31, 43; limitation of, 43, 45n
Fishlow, Albert, 102, 103n
Fonds Europeen d'Orientation et de Garantie Agricole (FEOGA)
Food: political economy of, 1–3; as weapon in foreign relations, 25, 223–25

Food and Agriculture Organization (FAO), 37, 167, 183, 191–98, 201–4, 217; committees and working groups, 192–93; Indicative Plan, 1969, 195, 201, 202
Food policy (agricultural policy): Africa, 177–86, 220, 223, 232, 233; China, 110, 118–20, 122–26, 132, 216, 217, 221; Eastern Europe, 48–49, 65–77, 214–15, 232–33; Europe, 30–47, 214, 223; Latin America, 89–92, 217, 220; Middle East, 216, 217, 223, 232, 233; natural limitations, 209, 212, *table*, 210–11; North America, 11–29, 214, 220; policy recommendations, 228–36; regional approach to, 1–8, 207–37; Soviet Union, 48–80, 200, 214–15, 217, 232–33; summary, 209, 213–17, *table*, 210–11
Forman, Shepard, 103n, 105n
Fowke, Vernon C., 13
Fraenkel, Richard M., 165n
France, 223; agricultural interest groups, 47n; Mediterranean farm crops, 41–43; Polish trade with, 71, 73
Freedom from Hunger Campaign, 193–96, 202
Freyre, Gilberto, 99, 103n
Fukui, Haruhiro, 139n

Gabeli, Mustafa al-, 164n
Ganzel, Richard, 7, 220, 221, 222, 230, 236n
General Agreement on Trade and Tariffs (GATT), 2, 4, 27, 31, 32, 34, 35, 37–40, 45, 223
George, Susan, 236n
German Democratic Republic (East Germany), 68, 69, 71, 74; trade with German Federal Republic, 74, 75; USSR exports to, 59
German Federal Republic (West Germany), 74, 75
Ghedira, M., 198, 205n, 206n
Gierek, Edward, 69
Gil, Federico, 104n
Goldberg, Peter A., 104n
Goldman, Marshall, 78n
Goldwert, Marvin, 97, 104n
Gomulka, Wladyslaw, 69
Goulart, João, 100, 101
Great Britain: rural districts, 31; trade with Poland, 73
Greece, 37, 38, 41, 42, 44
Green forces in Europe, 42, 44, 47n
Green Revolution, 127, 144, 201, 236n
Griffin, Keith, 139n
Grunwald, Joseph, 90, 103n, 104n, 105n
Gundelach, Finn, 38

Hager, Wolfgang, 47n
Haiti, 89
Hardin, Garrett, 89, 103n
Harris, Simon, 46n, 47n
Harrison, Reginald J., 8n
Hayami, Yujiro, 134, 139n, 141n
Henshall, Janet D., 105n
Hermes, Peter, 80n
Herring, Ronald J., 157, 165n
Hill, Polly, 187n, 188n
Hoffman, George, 79n
Hong Kong, 110
Hopkins, Raymond, 8n, 29n, 84, 89, 103n, 191, 197–98, 203, 204–6n
Hua Guofeng, 124, 125
Hungary, 65, 67, 68, 73, 76; land ownership, 69–71; Technologically Operated Production System (TOPS), 71–72
Hunter, Guy, 187n

Iberia, 37, 38, 42, 43 (*see also* Portugal; Spain)
Indonesia, 126, 130, 131, 132, 135, 137; Chinese in, 184; food production and imports, 127, 216; land reform, 129; Indus River Basin, 144
Intergovernmental Maritime Consultative Organization (IMCO), 192
International Atomic Energy Agency (IAEA), 192
International Bank for Economic Cooperation, 74
International Fund for Agricultural Development, 193, 195–96, 198
International Monetary Fund, 34
International organizations, 191–206, 217, 225–28; improbability of global regime, 198–203
International Wheat Agreements, 202
Iran, 163; administrative structure, 153; agrarian reform, 158; agricultural extension services, 154; agriculture, 144, 148; cooperatives, 157
Iraq: agrarian reform, 158; agriculture, 144, 148; cooperatives, 157
Ireland, 42; fisheries, 31
Irigoyen, Hipólito, 96
Israel, agriculture, 144, 147, 148
Italy, farm crops, 41–43

Jackson, Roy, 205n
Jamgotch, Nish, 80n
Japan, 107, 114–18, 133, 137; agricultural cooperatives, 117; agricultural policies, 114–18, 213, 223; Canadian exports to, 24; currency, 116; European Community relations with, 38; food production and prices,
114, 130–31, *table*, 114; industrial development, 110, 131; interdependence, 32; Liberal Democratic Party, 117; Meiji restoration, 114–15, 117, 131; political situation, 116–18; population and income growth, 110; rice production, 115–17; trade barriers, 117; trade policies, 221, 222; U.S. embargo on soybean exports to, 40, 77, 223; U.S. relations with, 38, 40
Johnson, Chalmers, 141n
Johnson, D. Gale, 28n
Johnson, Harry, 139n
Johnston, B. J., 188n
Johnston, Bruce, 138–39n
Jones, William, 176, 187n, 188n
Jordan, Robert, 204n, 205n
Jordan, 144, 148, 163, 164
Josling, Timothy, 47n

Kapstein, Jonathan, 187n
Karcz, Jerzy F., 78n, 79n
Karlik, John, 47n
Karnow, Stanley, 140n
Kaufman, Robert D., 104n
Kautsky, Karl, 77n
Kazakhstan, 54
Kennedy Round negotiations, 38, 39
Kenya, 173, 179, 180
Keohane, Robert O., 8n, 20, 21, 24, 29n, 237n
Keynes, John Maynard, 201
Khan, Dilwar Ali, 165n
Khrushchev, Nikita, 53–55, 57, 64, 68
Kilby, Peter, 138–39n, 188n
Kilpatrick, James, 120, 140n
Knorr, Klaus, 8n, 45n
Korbanski, Andrzej, 80n
Korea (*see* North Korea; South Korea)
Kosygin, Aleksei, 55, 78n
Kubitschek, Juscelino, 100
Kuwait, 146

La Follette, Hugh, 237n
Laird, Betty A., 78n, 183, 188n
Laird, Roy D., 78n, 183, 188n
Laos, 118, 119, 131, 209
Lappé, Frances Moore, 237n
Lardinois, Pierre, 46n
Latin America, 83–105, 230; food and agricultural policy, 89–92, 217, 220; Free Trade Area, 91; *latifundia* system, 86, 87, 99
Lebanon, agriculture, 144, 147, 148
Leff, Nathaniel H., 105n
Leiden, Carl, 165n
Leigh, Michael, 47n
Lenin, Nikolai, 51–52, 64
Levi, Werner, 8n

Lewin, M., 78n
Libya, 148, 158, 163
Liberman, E., 55, 68
Lipsky, Seth, 141n
Liu Shaoqui, 122
Lomé Agreement, 233
Lopez Portillo, José, 20
Lovell, C. A. Knox, 78n
Loveman, Brian, 93, 103n, 104n
Lowi, Theodore J., 28n, 29

Macfarlane, David L., 22, 29n
Makings, S. M., 186n
Malaysia, 126, 127, 129–30, 132, 135, 136, 137, 216
Mallon, Richard D., 104–5n
Malloy, James M., 105n
Mamalakis, Markos, 104n
Manning, Travis W., 28n
Mao Tse-tung, 122–24, 132, 140n
Marcos, Ferdinand, 129
Mares, Vaclav E., 80n
Marsh, John, 46n
Mediterranean areas: European relations with, 33, 34, 37, 40–45; NATO, US and USSR in, 44
Mexico, 92; agrarian protests, 13; *Compañia Nacional de Sussistencias Populares*, 16; economic relations with US, 11; *ejidal* (communal agriculture), 12–13; Export Bank (*Banco de Comercio Exterior*), 25; farm products, reserve policy, 20; foreign trade policy, 15; *hacienda* agriculture, 13; interdependence with US, 20–23; laborers from, in US, 18; as less developed country, 20; oil resources, 14; *Partida Revolucionario Institucional*, 14; trade liberalization, 23–24
Middle East, 143–65, 209 (*see also* Egypt; Iran; Pakistan; other countries); agrarian reform, 157–60, 162–63; agricultural development, 143–49, 152–53, 232, *table*, 146; agricultural extension services, 154–55; agricultural policies, 216, 217, 223, 232, 233; arable land, 145–46, *table*, 146; bureaucracies, 149–53, 156; cooperatives, 156–57; credit institutions, 155–56; deserts, 145; food production, 144–49, *tables*, 145, 146; foreign experts in, 152; implementation of government policies, 153–57; industrialization, 144; NATO in, 44; policy pressures and styles, 160–63; population expansion, 144; regional integration, 164; research in agriculture, 151–52; US in, 44; USSR in, 44
Millar, James, 78n
Miracle, M. P., 188n

Momsen, R. P., 105n
Moore, Richard, 6–7, 208, 214, 220, 221, 222, 227, 234, 236n
Morocco, 144, 148, 150
Morris, Cynthia, 139n
Morse, Edward S., 8n, 45n, 46n
Musgrave, Philip, 90, 103n, 104n, 105n

Nau, Henry, 205n
Neal, Fred W., 79n
New Zealand, 37
Nicholson, Norman K., 88, 103n, 165n
Nigeria, 173
North (developed countries), 6
North America, 11–29 (*see also* Canada; Mexico; United States); food policy, 11–29, 214, 220; regional interdependence, 20–28, 227, 228; trade liberalization, 23–24
North Atlantic Treaty Organization (NATO), 42, 44
North Korea, 118, 119, 131
Nove, Alec, 78n
Nurske, Ragnar, 103n
Nye, Joseph S., 8n, 20, 21, 24, 29n, 237n

Office of Economic Cooperation and Development (OECD), 37
Ohlin, G., 187n
Oil producing and exporting countries (OPEC), 130, 193, 223
Oil resources: Brazil, 100, 101; conservation in Europe, 44; Mexico, 14; Southeast Asia, 136
Okwuosa, Emmanuel, 182, 185, 186n, 188n
Organization of Economic Cooperation and Development (OECD), 14, 17, 130
Osborn, Robert, 77n
Osborne, R. H., 79n
Overhold, William, 141n

Paarlberg, Robert, 29n, 139n, 236n
Pakistan, 144, 150, 163; agrarian reform, 158; agricultural extension services, 154, 155; agriculture, 144–47; arable land, 145–46; food imports, 147; government departments, 153
Paraguay, 91
Patterson, Kathleen, 45n
Paz, Octavio, 22, 29n
Pearson, Lester B., 205n
Pentland, Charles, 8n
Perón, Juan Domingo, 97, 98, 105n
Peterson, Trudy Huskamp, 80n
Petras, James, 94, 104n
Philippines, 126, 129–30, 135, 136, 138
Piekalkiewicz, Jaroslaw A., 79n
Pike, Frederick B., 105n
Pirages, Dennis, 28n, 103n
Plaxico, James, 45n

Ploss, Sidney I., 78n
Poland, 65, 67, 68, 69, 71, 76; agricultural imports, 73; credit from foreign nations, 73; land ownership, 67, 69; U.S. aid and trade relationship, 74; USSR exports to, 59
Portugal, 38, 41–44
Pospielovsky, Dimitry, 78n
PRC (see China)
Preston, L. E., 188n
Puchala, Donald, 8n, 29n, 84, 89, 103n, 191, 197–98, 203, 204–6n
Pura, Raphael, 141n

Quadros, President (Brazil), 100

Race, Jeffrey, 142n
Randall, Laura, 105n
Red Cross, International, 217
Reutlinger, Sholmo, 139n, 141n
Reynolds, Bruce, 140n
Reynolds, Clark W., 104n
Richter-Altschaffer, J. H., 28n
Rigby, T. H., 77n
Robbins, Roy M., 28n
Robbins, Williams, 79n
Roca-Runciman Treaty, 96
Rock, David, 104n
Roett, Riordan, 105n
Romanowsky, Jacek I., 79n
Rourke, Francis F., 29n
Rumania, 65–69, 71, 73, 74
Ruttan, Vernon, 134, 139n, 141n

Sachs, Moshe, 204n, 205n
Salazar-Carillo, Jorge, 90–92, 104n
Saleh, Abdullah, 199, 205n
Saouma, Director-General (FAO), 196, 200–201
Sarmiento, Domingo, 104n
Sarris, Alexander, 206n
Saudi Arabia: agricultural imports, 148; aid to Sudan, 146–47
Savary, R., 205n
Scarlett, Lynn, 7–8, 220, 221, 230, 231
Scheffer, Paul, 77n
Schnittker, John A., 28n
Schoonover, David, 78n
Schultz, T. Paul, 187n
Scobie, James R., 104n
Seevers, Gary L., 29n
Selowsky, Marcelo, 139n, 141n
Sen, B. R., 193
Shaffer, Harry, 77n
Shee Poon-Kim, 142n
Shepherd, Jack, 236n
Shlaim, Avi, 47n

Shokolov, Mikhail, 78n
Shrader, W. D. 186n
Shue, Henry, 28n, 29n, 236n
Shuman, Charles B., 28n, 29n
Sierra Leone, 173, 180
Sigmund, Paul E., 104n
Sindhwani, Trilok N., 139n, 142n
Singapore, 135, 136, 137; population and income growth, 110
Singlemann, Joachim, 141n
Skidmore, Thomas E., 105n
Skinner, G. William, 141n
Smith, Peter, 104n
Sourrouille, Juan V., 105n
South (underdeveloped regions), 6
South America (see Latin America)
South Korea, 110
Soviet Union (USSR), 6, 68, 209; Afghanistan invaded by, 62, 74, 225; agricultural policy, 48–80, 200, 214–15, 217, 232–33; collectivization, 52–53; grain exports, 59, 71, table, 58; grain imports, 59, 62, 77, 223, 234, tables, 60, 61, 63; grain production, 57, table, 50; integration with Eastern Europe, 226–27; kulaks, 52, 53, 77n; Machine Tractor Stations (MTS), 52–54; in Mediterranean and Middle East region, 44; New Economic Policy (NEP), 51–52; population and GNP rates, table, 56; relations with China, 120, 126, 135; relations with Southeast Asia, 135; U.S. grain embargo against, 24, 25, 48, 49, 62, 64, 75, 77, 223–25, 234, 236–37n; U.S. grain exports to, 59, 62, 198, 223, 224, tables, 60, 63
Spain, 38, 41, 44
Springborg, Robert, 165n
Stalin, Joseph, 51–53, 64, 68, 75, 78n
Stavis, Benedict, 140n
Stein, Barbara H., 99, 105n
Stein, Sidney J., 99, 105n
Stepan, Alfred, 105n
Stritch, Thomas, 105n
Sudan, 144–47; arable land, 145–46; food imports, 147; land reform, 158
Suharto, President (Indonesia), 127, 130, 135
Swinbank, Dan, 47n
Syria, 164; agrarian reform, 158; agriculture, 148; cooperatives, 157
Syvrud, Donald E., 105n

Taiwan, 110, 131
Tanzania, 180, 181
Taylor, Lance, 206n
Thailand, 126, 127, 129–30, 132, 134, 135, 136, 138, 216

Thiesenhusen, William C., 93, 104n
Thompson, Louis, 186n
Thompson, Seth, 8, 208, 217
Tigris-Euphrates river system, 144
Trager, James, 29n
Treaty of Rome, 37
Trethaway, Richard J., 28n
Tsoukalis, Loukas, 47n
Tucker, Robert W., 87–88, 103n
Turkey, 144, 148
Turley, William S., 142
Tweeter, Luther, 45n
Twitchett, Carol Cosgrove, 46n

Uchendu, Victor, 188n
Uganda, 174, 180; Asians in, 184
Ul Haq, Mahbub, 237n
United Kingdom (see Great Britain)
United Nations Commission for Trade and Development (UNCTAD), 223
United Nations Conference on Food and Agriculture, 1943, 192
United Nations Development Program, 193, 194, 204
United Nations Economic and Social Commission for Asia and the Pacific (ESCAP), 135
United Nations Educational, Scientific and Cultural Organization (UNESCO), 192
United States: Agricultural Act (1954), 16; Agricultural Adjustment Act (AAA, 1933) 17–18; agricultural exports to USSR and Eastern Europe, 73, table, 72; agricultural policy, 11–29, 214, 220, 222–23; agricultural trade, 36, 40, 214, 220, 222–25; Agricultural Trade Development and Assistance Act, 222; Agricultural Trade Expansion Act, 73; Agriculture and Consumer Protection Act, 18; Commodity Credit Corporation, 19, 22, 25, 62, 73; consumer movement, 19; consumer protection, 18; credit to Eastern Europe, 73; embargo on soybean exports to Japan, 40, 77, 223; European relations with, 33, 34, 37–40; farm income maintenance, 12, 18; farm market control, 200; farm products, parity system, 17–18; farm products, prices, 15–18; farm products, reserve policy, 19–20; Food and Agriculture Act (1965), 18; Food for Peace program, 71; Foreign Agricultural Service, 62; grain embargo against USSR, 24, 25, 48, 49, 62, 64, 75, 77, 223–25, 234, 236–37n; interdependence with Canada and Mexico, 20–23; interdependence with Europe, 32; investments in Southeast Asia, 138; Japanese relations with, 38, 40; labor supply, 18–19; in Mediterranean and Middle Eastern region, 44; New Deal, 13, 19; Polish aid and trade relationship, 74; trade embargo against Eastern Europe, 71, 74; trade liberalization, 23–24; trade with USSR, 59, 62, 223; USSR grain imports from, 59, 62, table, 60
Uruguay, 91
USSR (see Soviet Union)

Valenzuela, Arturo, 104n
Vargas, Getulio, 99–101, 105n
Vastine, John R., 46n
Venezuela, 91
Vietnam, 118, 126, 131, 135
Viner, Jacob, 8n
Volgyes, Ivan, 79n, 80n
Volle, Angelika, 47n

Wadëkin, Karl-Eugen, 79n
Walker, Kenneth R., 141n
Wallace, William, 47n
Wanandi, Jusuf, 142n
Warley, Thorald K., 16–17, 28n, 45n
Waterbury, John, 164n, 165n
Watt, D. C., 47n
Weinbaum, Marvin G., 7, 164n, 165n, 217, 230, 236n
Weiss, Thomas, 204n, 205n
Wen Rong, 140n
Whetham, Edith, 181, 187n, 188n
Wong, John, 142n
World Bank, 193, 204
World Food Conference, Rome, 1974, 4, 191, 193, 195–96, 204
World Food Congresses, 1963 and 1970, 191
World Food Council, 193, 195, 198, 203
World Food Program, 193–97, 200, 201, 202
World Health Organization, 192
World Meteorological Organization (WMO), 192
Wortman, Sterling, 237n
Wynia, Gary W., 85, 103n, 105n

Yemen PDR, 144, 158
Yergin, Daniel, 78n
Yugoslavia, 68, 71, 74; credit from foreign nations, 73; land ownership, 67, 69

Zaire, 174
Zambia, 180
Zeller, Adrien, 46n
Zhou Enlai, 124
Zysman, John, 28n

Contributors

DAVID N. BALAAM, assistant professor of political science at the University of Puget Sound, Tacoma, Washington, is the author of "American Food Policies and Strategies: Can the Hunger Problem Be Managed?" in John Peters, ed., *Issues in Agricultural Politics and Policy*. Current research: international food production and distribution trends.

MICHAEL J. CAREY, assistant professor of political science, Loyola Marymount University, Los Angeles, is the author of "Parties and the Irish Political System," in Peter Merkl, ed., *Western European Party Systems* and "Catholicism and Irish National Identity," forthcoming. Current research: fisheries policy.

ROBERT E. DICKENS, Director of Teleconference Instruction, University of Nevada, Reno, formerly taught political science at California State University, Fullerton, and at UNR. Current research: natural resources policy.

RICHARD GANZEL, associate professor of political science, University of Nevada, Reno, is the author of various studies on energy, resource policy, and geothermal energy, and coeditor of *Energy and Nevada*.

RICHARD K. MOORE is an assistant professor of political science, Lewis Clark State College, Lewiston, Idaho. Current research: international agri-policy making, Northwest farm exports.

LYNN SCARLETT is a doctoral candidate at the University of California, Santa Barbara. Current research: technology and international politics.

SETH B. THOMPSON is an associate professor of political science, Loyola Marymount University. Current research: delegation behavior in the U.N. General Assembly, and the application of attribution theory to political behavior.

MARVIN G. WEINBAUM, an associate professor of political science, University of Illinois, Urbana-Champaign, has had articles on policy elites in the Middle East published in *The Middle East Journal*, *Asian Survey*, and *Studies in Comparative International Development*. Current research: forthcoming book on food and development in the Middle East.